Ultrafine-Grain Steels: Mechanical Behavior

Ultrafine-Grain Steels: Mechanical Behavior

MARIA JOSE QUINTANA

ROBERTO GONZALEZ

Printed by CreateSpace, An Amazon.com Company

Available from Amazon.com and other book stores

Cover by Regina Meza

ISBN-10: 1530473675
ISBN-13: 978-1530473670

Preface

Development of new materials for transportation is usually related to composites and light alloys, however the automotive sector (as well as most metal-mechanical transformation industries) is still researching in the advancement of modified and more efficient steel parts. Strength and performance of high-alloy steels is a possible solution for many design applications, yet, the cost-efficiency generally sought in this industrial sector has promoted the use of low-alloy and high-strength steels, specially in the form of sheets to produce stamped parts.

This type of high-strength low-alloy steels are produced by advanced thermomechanical controlled rolling processes and are in fact ultrafine grained steels with grain sizes below 5 μm, resulting in parameters such as coefficient of strain hardening, mechanical strength and admissible thickness tolerances in steel sheets, in order to use these materials in subsequent cold-work operations (bending and drawing).

In recent years, both the steelmaking industry and laboratories in different parts of the world, have shown an increasing interest in reaching an industrial-level production of ultrafine grained steels (also known as ultrafine ferrite), which have a grain size d lower than 5 μm, and enhanced mechanical resistance and fracture toughness. Though different laboratory techniques such as severe plastic deformation, accumulative roll bonding, dynamic strain induced deformation and rapid transformation annealing, have produced utrafine grained steels, plastic instability consequence of the low strain hardening n coefficient prevents industrial production. Only advanced rolling techniques such as advanced thermomechanical cold rolling processes have produced industrial level amounts of steel sheets with mechanical and commercial characteristics.

The production of high-strength low-alloy or dual-phase steels with strict control of parameters of thermomechanical rolling processes and cooling rates to form specific phases results in ultrafine-grained materials with high mechanical resistance, high impact strength, and very good formability.

Thermomechanical hot rolling processes are done at temperatures above Ar_3, with strict control of rolling parameters and cooling and coiling rates to allow the manufacture of finished products with grain sizes as small as 1 μm. These steels show a high strain hardening coefficient value (this coefficient must have a value higher than 0.1 in order to allow the stamping of final parts), unlike commercial high strength low alloy steels manufactured by other similar processes. This procedure generally requires the addition of Si, Cr, Mo and Mn combined with rapid cooling of the strip after the final rolling (finishing) pass, followed by coiling after which a controlled cooling rate of must be maintained. Under these conditions, the structure is formed by polygonal ferrite while avoiding pearlite. Furthermore, the remaining islands of carbon-enriched austenite may or may not transform into bainite–ferrite during cooling of the coiled strip.

At room temperature, the capacity of the material to be deformed during bending or drawing operations (typical of requirements for automotive parts applications) depends on the interaction of a hard and a soft phase in the microstructure. On the other hand, at high temperatures, these steels may show superplastic behavior if deformed at a precise combination of temperature and strain rate. The Ashby–Verrall model provides a satisfactory description of the mechanisms of grain boundary sliding and dislocation creep that occur during superplastic deformation.

Contents

List of figures and tables

Symbols
and Formulae

ASTM G	Grain size (ASTM standard)
$\boldsymbol{b}$	Burgers' vector (μm)
d	Grain size (μm)
D	Grain size of the matrix (μm)
f_v	Volume fraction of the precipitate phase
ITT	Impact transition temperature (°C, K)
m	Strain-rate sensitivity exponent
n	Strain hardening coefficient
T_M	Melting temperature (°C, K)
T_{nr}	Non-recrystallization temperature (°C, K)
r	Radius of the precipitate (μm)
γ	Interfacial energy (J/m^2)
$\dot{\varepsilon}$	Strain rate (s^{-1})
σ	Stress (MPa)
σ_y	Yield stress (MPa)

Stress - strain

If stress is a function of strain:

$$\sigma = K\varepsilon^n$$

n = strain hardening coefficient, K = constant. To obtain the strain hardening coefficient, the same equation can be expressed as:

$$\log \sigma = n \log \varepsilon + \log K$$

If stress is a function of the strain rate, then:

$$\sigma = K\dot{\varepsilon}^m$$

m = strain-rate sensitivity exponent, and can be experimentally obtained by:

$$m = \left(\frac{\log(\sigma_{y_2}/\sigma_{y_1})}{\log(\dot{\varepsilon}_{0_2}/\dot{\varepsilon}_{0_1})} \right)_{T,d,\varepsilon}$$

σ_y = the yield stress at 0.2%, $\dot{\varepsilon}_0$ = initial strain rate, in tests made at two different strain rates.

Yield stress, ITT and grain size

$$\sigma_y \approx 54 + 17d^{-1/2}$$

$$ITT \approx -19 - 11.5\, d^{-1/2} + 2.2\ (\text{pearlite } \%)$$

σ_y = yield stress, d = grain size, ITT = Impact transition temperature.

Strain rate and grain size

$$\dot{\varepsilon} \approx (\mathbf{b}/d)^p$$

$\dot{\varepsilon}$ = strain rate, $\mathbf{b}$ = Burgers vector, $p = 2$ when the process is related to lattice-diffusion-controlled creep (Nabarro–Herring creep) or $p = 3$ when it is related to grain-boundary diffusion-controlled creep (Coble creep).

Zener formula

$$D \cong \frac{2r}{f_v}$$

D = grain size of the matrix, f_v = volume fraction of the precipitate phase, r = precipitate radius.

Pulling and pinning force of the precipitates

Pulling or traction force:

$$\sim 2\gamma/\bar{D}$$

γ = interfacial energy, D = grain size of the matrix.

Pinning force

$$\frac{3f\gamma}{2r}$$

f = volume fraction of precipitates, γ = interfacial energy, r = precipitate radius.

Solubility of precipitates

Titanium nitride

$$\log[N][Ti] = \frac{-14400}{T} + 5$$

$[N]$ = amount of nitrogen, $[Ti]$ = amount of titanium, T = temperature (Kelvin).

Titanium carbide

$$\log[C][Ti] = \frac{-7000}{T} + 2.75$$

$[C]$ = amount of carbon.

Niobium carbide

$$\log[C]^{0.87} [Nb] = \frac{-7530}{T} + 3.11$$

$[Nb]$ = amount of niobium.

Niobium carbonitride

$$\log[N]^{0.65} [C]^{0.24} [Nb] = \frac{-10400}{T} + 4.09$$

Power-law creep, or Weertman creep

$$\dot{\varepsilon} = A\sigma^n \exp\left(-\frac{Q}{RT}\right)$$

A, n = constants, σ = stress, Q = activation energy of creep ($\approx$self-diffusion), R = the ideal gas constant, T = absolute temperature.

Exponential-law creep, or Dorn creep

$$\dot{\varepsilon} = A \sinh(\beta\sigma) \exp\left(-\frac{Q}{RT}\right)$$

A, β = constants.

Grain boundary diffusion

The diffusion coefficient at intermediate temperatures $(0.3T_M < T < 0.5T_M)$:

$$D_{it} \approx D_v\left(1 + \frac{\delta}{d}\frac{D_{gb}}{D_v}\right)$$

D_v = volume diffusion coefficient, D_{gb} = grain boundary diffusion coefficient, δ = width of the grain boundary ($\approx$ two interatomic distances, $2b$).

Herring–Nabarro creep

$$\dot{\varepsilon} = \frac{B\sigma \exp(-Q/RT)}{d^2}$$

B = constant.

Ashby–Verrall equation for superplasticity

$$\dot{\varepsilon}_{\text{total}} = \dot{\varepsilon}_{\text{D-A flow}} + \dot{\varepsilon}_{\text{dislocation creep}}$$

$\dot{\varepsilon}_{\text{D-A flow}}$ = diffusion-accommodated flow, $\dot{\varepsilon}_{\text{dislocation creep}}$ = plastic flow caused by creep due to the movement and climbing of dislocations.

$$\dot{\varepsilon}_{\text{D-A flow}} = 98\frac{\Omega\mu}{kTd^2}\left(\frac{\sigma}{\mu} - \frac{0.72\,\Gamma}{\mu d}\right)D_v\left(1 + \frac{\pi\delta}{d}\frac{D_B}{D_v}\right)$$

$$\dot{\varepsilon}_{\text{dislocation creep}} = A_1\frac{\mu b}{kT}\left(\frac{\sigma}{\mu}\right)^n \exp\left(-\frac{Q_c}{RT}\right)$$

Ω = atomic volume, μ = shear modulus, k = Boltzmann's constant, T = absolute temperature, Γ = grain boundary free energy, D_v = bulk diffusion coefficient, δ = thickness of the boundary, D_B = boundary diffusion coefficient, A_1, n = constants, b = Burgers' vector, Q_c = activation energy for dislocation creep, R = ideal gas constant.

1
Ultrafine-Grain Steels

Though the development of advanced materials (also known as new materials), information technologies, artificial intelligence, and biotechnology are considered to be the most important industries for the 21st century[1], price reductions and improvements in quality and properties in steel have maintained this material as a leader in all industrial applications[2]. More specifically, the use of structural materials in the second half of the 20th century included the competition between Fe-C alloys and other metallic materials, which resulted in advanced steels that respond to large production necessities. The interest in Fe-alloys is based, among other things, in its changes during solid-state transformations, restoration and recrystallization, and textures[3].

For example, there is an ongoing activity in the steel industry to develop new methods to produce high-strength low-alloy (HSLA) structural steels with lower cost and improved properties. The use of heat treatments such as normalizing and quenching and tempering, and more recently, thermomechanical controlled rolling processing techniques (TMCRP) and the use of continuous annealing processing lines (CAPL) have been developed to produce fine ferrite grain sizes in final products. These techniques have shown a significant improvement in strength, fracture toughness, and weldability through the refinement of ferrite grain size (up to 12 ASTM G, $d_\alpha \approx 5$ μm), which used to be the lowest grain size reachable in the industry 30 years ago[4]. However, the Hall–Petch equation predicts that a reduction from 5 to 1 μm should increase the yield strength of a given steel up to 350 MPa and decrease the impact transition temperature (ITT) to -100°C[5]. The use of current commercial techniques like TMCRP, which improve properties and allow the replacement of some costly alloy steels with either plain carbon or microalloyed/low-alloyed carbon steels, demonstrate that the industrial production of fine equiaxed ferrite grains with $d = 2\sim3$ µm, is possible.

Thermomechanical processing such as controlled rolling, controlled cooling and direct quenching save energy in steel manufacture by minimizing or even eliminating the heat treatment after hot deformation, thereby increasing the productivity of high grade steels. It generally demands a change in alloy design and frequently reduces the productivity of the hot deformation process itself, but makes it possible to reduce the total amount of alloying additions and to improve strength, toughness, and weldability, whilst sometimes producing new and beneficial characteristics in the steel.

A large amount of the rolled steel products currently produced are of the microalloyed type. These steels are usually soaked at high temperatures when roughening deformation is carried out. In the case of conventional controlled rolling (CCR), rough rolling is followed by fast cooling and finishing passes are carried out at temperatures where the austenite remains un-recrystallized. The microalloying elements, which remain in solution (partially or completely) during rough deformation, start to precipitate after finishing deformations at low temperatures. Nb, Ti and V are the most commonly used microalloying elements and during cooling, they combine with C and/or N to form carbide, nitride and/or carbonitride precipitates. These fine precipitates play an effective role by retarding recrystallization (and therefore, increasing the recrystallization-stop temperature) that usually follows deformation and thus, helps to retain the accumulated strain and deformed structures of austenite grains.

1.1 Controlled Rolling Processes

The manufacture of dual-phase (DP) steels by Advanced Thermomechanical Controlled Rolling Processes (ATMCRP), using temperatures close to A_{r3} (the temperature at which, during cooling, the austenite-to-ferrite ($\gamma \to \alpha$) transformation starts) involves strict control of the rolling process variables, as well as the cooling rates, in order to allow the formation of ultrafine-grain (UFG) materials with grain size close to 1 µm[6; 7; 8].

Through direct quenching (by a fine spray of water over the hot material), the amount of martensite present in DP steels can be modified, as well as the size and distribution of ferrite (soft phase), producing finished products with tension stresses as high as 800 MPa[8].

These processes include careful control of chemical composition (including microalloying elements such as Ti, Nb and V), deformation sequences, austenitic non-recrystallization temperatures, and the $\gamma \to \alpha$ allotropic transformation temperature during cooling[9]. In laboratory experiments, for certain combinations of these parameters, various severe plastic deformation (SPD) processes can be used to produce UFG materials, but these techniques have not yet reached industrial-scale production. The following are examples of SPD processes:

- accumulative roll-bonding (ARB)[10]
- dynamic strain-induced transformation (DSIT)[11]
- cold rolling and annealing (CRA)[12]
- rapid transformation annealing (RTA)[13]
- equal-channel angular pressing (ECAP)[14]
- high-pressure torsion (HPT)[15; 16]

ATMCRP applied to HSLA steels improve their properties and reduce the costs of parts manufactured with these materials. They have greater strength, fracture toughness, and weldability because of the refinement of ferrite grain size in the final products[4; 9].

Thermomechanical processing minimizes and/or eliminates the need for heat treatment after hot deformation, reducing costs and increasing productivity. Moreover, it allows the production of newer materials, such as steels with lower amounts of alloying elements[9].

According to some authors, the manufacture of DP steels by ATMCRP involves four steps as shown in Figure 1.1[17]:

1. Rolling in roughening and finishing mills, refining the austenite grain size through repeated static recrystallization as well as deformation of austenite in the non-recrystallization region.
2. Water cooling.
3. Isothermal holding at intercritical temperatures.
4. Rapid continuous cooling to the required coiling temperature. A martensitic transformation takes place if the presence of bainite is undesired.

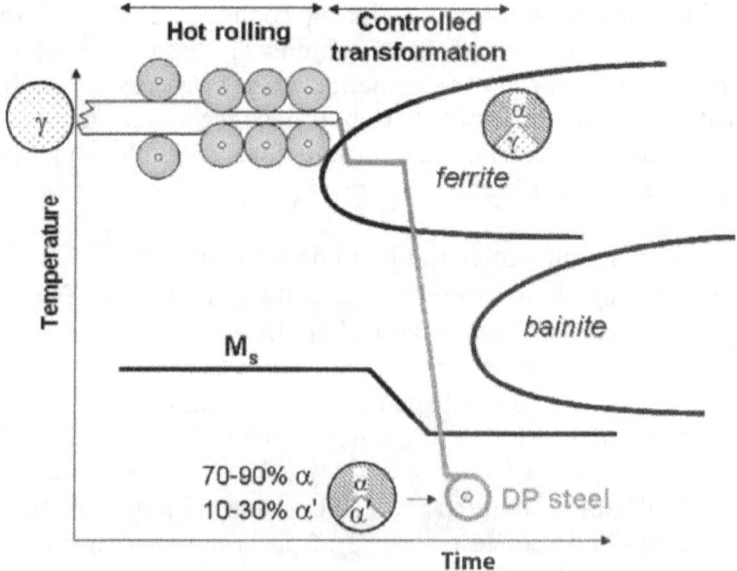

Figure 1.1 Sketch of the processing of DP steels [17]

Other authors consider ATMCRP as involving six steps, in order to obtain a sufficiently high value of the strain hardening coefficient n to allow component production[8]:

1. *Homogenization.* In order to start with a recrystallized structure, the continuous casting slabs (200~250 mm) are kept at 1200~1250°C.
2. *Roughening.* In order to reduce the thickness (~20 mm), approximately 10 passes are made in reversible rolling mills at 1100~1200°C.
3. *Waiting.* The material is cooled to 1000~1100°C.
4. *Finishing.* The thickness is reduced (~1.5 mm) through a mill of hot or semi-continuous bands, while the temperature drops to 850°C.
5. *Controlled cooling.* In order to reach the coiling temperature (~600°C) the steel is cooled by a fine water spray. At this step, a partial $\gamma \rightarrow \alpha$ transformation takes place; if the cooling rate increases, the ferrite percentage decreases.
6. *Coiling.* The material is coiled at ~600°C. Below this temperature, the untransformed austenite becomes bainite–martensite (even if the steel is already coiled). The coiling window between the ferritic–pearlitic zone (upper) and the bainitic–martensitic zone (lower) must match this temperature.

Other authors propose three options for ATMCRP[18; 19]:

- *Option 1.* Starting from a reheated austenite (after the thermal treatment of homogenization), followed by successive straining and recrystallization at temperatures higher than A_{r3}, producing ferrite with grain sizes less than 10 μm.
- *Option 2.* Starting from the same austenite, which is successively deformed and recrystallized between the different rolling passes, except the final finishing passes, producing a non-recrystallized austenite that is then transformed into a very fine ferrite (with grain size less than 5 μm, 12 ASTM G).
- *Option 3.* This is equivalent to Option 2, but the finishing rolling is prolonged below the $\gamma \rightarrow \alpha$

transformation. In other words, strain and allotropic transformation may occur simultaneously in the last boxes of the finishing train (hot rolling mill), producing an ultrafine ferrite with a dual-phase microstructure and a mean grain size less than 5 µm (13 ASTM G).

Figure 1.2 shows a schematic representation of the three stages of ATMCRP and the microstructural changes due to deformation in each stage[9; 20]:

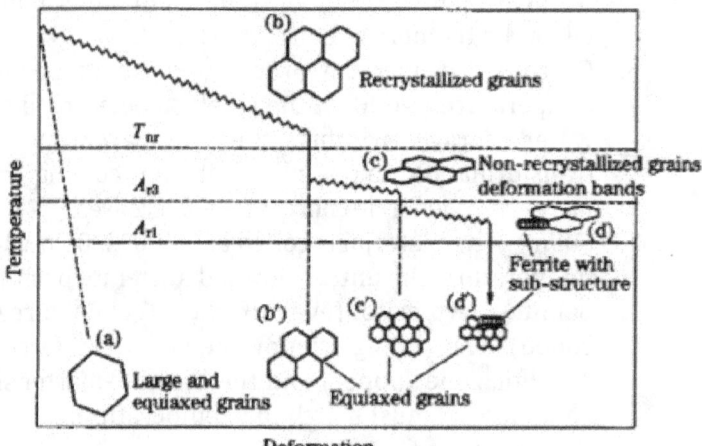

Figure 1.2 Schematic diagram of the controlled rolling process [9]

- *Stage 1*: refining coarse austenite (a) through repeated deformation and recrystallization (b). The steel transforms to relatively coarse ferrite (b′) except when microalloyed Ti is used, which allows refinement of homogenized austenite before rolling.
- *Stage 2*: forming elongated and non-recrystallized austenite bands (c) produced by deformation and nucleating ferrite both in the deformation bands and in the γ grain boundaries. This produces fine ferrite grains. Nb can be used to favor the accumulation of deformation inside the austenite

when the rolling process takes place below the non-recrystallization temperature T_{nr}. This temperature is a key factor to be controlled in ATMCRP.

- *Stage 3*: ongoing deformation in the ferrite–austenite dual-phase region, producing a substructure. If the steel is microalloyed with V, precipitation hardening by V_4C_3 will take place in the ferrite phase.

Sometimes there is a delay in the rolling between Stages 1 and 2, as well as accelerated cooling after Stage 3. This cooling along with deformation takes place in the $\gamma \rightarrow \alpha$ transformation temperature range just after the controlled rolling, refining the ferritic and martensitic grain sizes and thereby improving the strength and toughness of the steel. When the deformation takes place below the non-recrystallization temperature, the austenite grains become elongated and deformation bands are introduced inside the grains (pancaking). Therefore, the non-recrystallization temperature is the main controlling factor in ATMCRP[9; 20].

In the temperature range of interest for ATMCRP, the rates of diffusion of C and N are orders of magnitude greater than those of Ti and Nb. Therefore, Ti and Nb are the rate-controlling elements for these processes[9; 21; 22; 23].

In order to produce steels with ultrafine microstructures, careful control of chemical composition (microalloying elements such as Ti, Nb, and V), deformation sequences (ATMCRP), austenitic non-recrystallization temperatures, and $\gamma \rightarrow \alpha$ allotropic transformation temperature during cooling must be achieved during manufacture.

1.2 Microstructure of ultrafine-grain steels

The most common microalloying elements for HSLA steels are Ti, Nb, and V, while DP steels obtained from ATMCRP are also microalloyed with Mn, Si, Cr, Mo, and B. During the roughening processes, these materials are soaked at high temperatures and the microalloying elements remain in solution (partially or completely), only starting to precipitate later during the finishing deformations as the temperature falls. During cooling, the microalloying elements Nb, Ti, and V combine with C and/or N to produce carbide, nitride, and/or carbonitride precipitates. These hard particles retard the recrystallization that follows the deformation and help to retain the accumulated strain and deformed structures inside the austenite grains[9].

ATMCRP will produce elongated (pancake structure) grains in the rolling direction, which explains the elevated yield strength and ultimate tensile strength. This is specifically due to the amount of small ferrite grains (Figure 1.3) and independent of the amount of second-phase particles (martensite)[9].

However, grain size reduction does have a disadvantage: ductility as well as work-hardenability are lost in the process, accompanied by the "Lüdering effect" or "Lüders bands," which will lead to fracture. On the other hand, steels microalloyed with Ti, Nb, and V obtained by ATMCRP avoid this problem, since the stress/strain-induced $\gamma \to \alpha$ transformation favors work hardening and the anisotropy in the grains increases the number of interfaces per unit volume[9; 24].

Commercial ultrafine ferrite (UFF) steels obtained by ATMCRP are currently produced with grain sizes of 2~3 μm and yield stresses close to 700 MPa. These materials can be welded owing to their low carbon content and are therefore used in the automotive industry as construction and/or reinforcement parts, since they produce lighter vehicles without compromising safety standards[9].

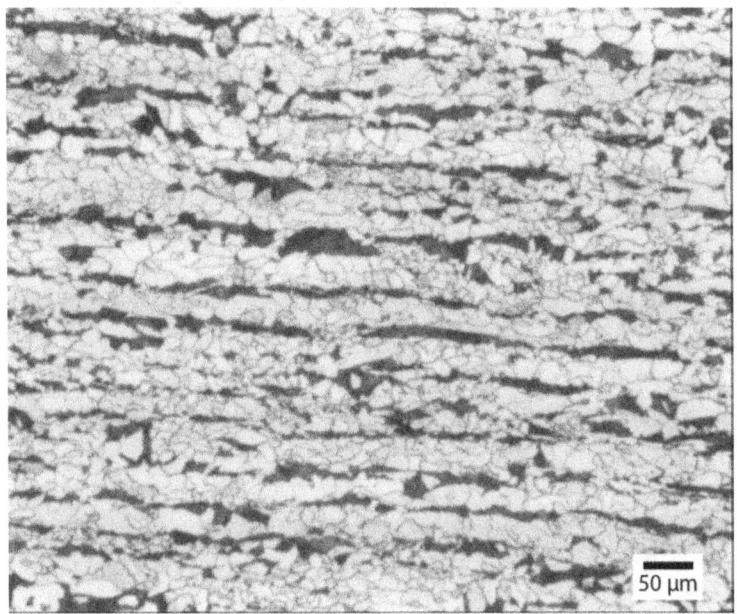

Figure 1.3 Microstructure of an HSLA steel obtained by ATMCRP

1.2.1 **Precipitates**

Microalloying elements, sometimes present themselves in the form of nitrides or carbides. The presence of titanium and niobium carbides (or carbonitrides) has a very important function on HSLA steels manufactured by ATMCRP processes, as follows:

- The titanium nitrides are responsible for maintaining the austenitic grain size during the homogenization process before rolling.
- The titanium nitrides and niobium carbonitrides act as an obstacle to the grain growth of the

recrystallized austenite in times between passes in the roughening process.

- The massive precipitation of niobium carbonitrides during the waiting time between the roughening and finishing mills, delay the static recrystallization of the austenite in the finishing process, allowing (after the allotropic transformation $\gamma \rightarrow \alpha$) the creation of fine and ultrafine ferrites (UFG steels).

- The niobium and titanium nanoprecipitates are responsible for an increase in the elastic limit, without a negative effect on the stiffness.

The classic Zener formula[25] shown in Figure 1.4, allows, in a semi-quantitative manner, equilibrium between the grain size of the matrix D, the volume fraction of the precipitate phase f_v and its grain size $2r$ according to the following expression:

$$D \cong \frac{2r}{f_v}$$

[1.1]

which balances the tendency of growth and the pinning effect produced by the precipitates. Thus, for precipitate volume fractions of $10^{-3} \sim 10^{-4}$ (as in the case of Ti and Nb precipitates) and grain sizes of the matrix (usually ferrite) between $1 \sim 10$ μm, the size of precipitates must be in the nanometric scale (between $1 \sim 10$ nm). Also, Figure 1.4 shows precipitates that are inhibiting the grain growth. These particles create a traction force (with an approximate value of $\sim 2\gamma/\overline{D}$) at the grain boundary and restricting its movement. However, there is an opposing force (pinning) to this traction force (pulling). When the mean grain size $\overline{D}$ increases, the pulling force decreases until it becomes insufficient and the grain growth stops.

Furthermore, it is necessary that, at the temperatures at which the precipitates are created (1200°C for titanium nitrides), the rolling process stops (delay time) in order for the precipitation to take place. Consequently, this type of rolling process is known as controlled rolling.

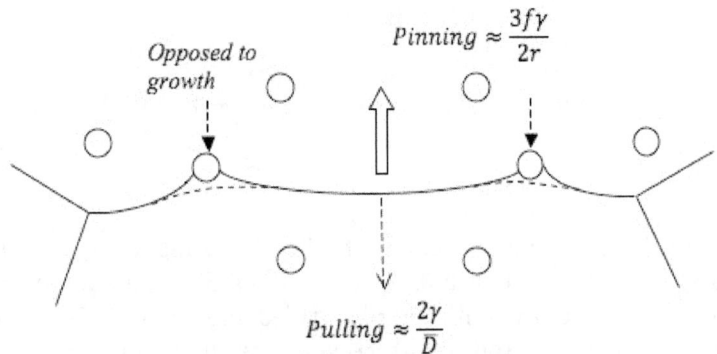

**Figure 1.4 Pinning effect: effect of precipitates
in the grain boundary migration**

The solubility formulas for the titanium nitride, titanium carbide, niobium carbide and niobium carbonitride are[26]:

- Titanium nitride

$$\log[N][Ti] = \frac{-14400}{T(K)} + 5$$

[1.2]

- Titanium carbide

$$\log[C][Ti] = \frac{-7000}{T(K)} + 2.75$$

[1.3]

- Niobium carbide

$$\log[C]^{0.87}[Nb] = \frac{-7530}{T(K)} + 3.11$$

[1.4]

- Niobium carbonitride

$$\log[N]^{0.65}\,[C]^{0.24}[Nb] = \frac{-10400}{T(K)} + 4.09$$

[1.5]

Considering a steel with the following composition: 0.168% C, 0.026% Ti, 0.033% Nb and 0.0055% N, the temperatures when the precipitation of nitrides and carbides begin to form can be calculated, and the results are shown in Table 1.1.

Table 1.1 Precipitation temperatures of titanium and niobium

Precipitate	T (K)	T (°C)
Titanium nitride	1628	1355
Titanium carbide	1370	1097
Niobium carbide	1430	1157
Niobium carbonitride	1439	1166

The rolling process (ATMCRP) consists in two stages: roughening and finishing; with a delay time between them where the niobium carbonitrides precipitation will take place. The previously mentioned carbides and carbonitrides will begin to precipitate in the austenitic phase therefore delaying both the grain growth and the recrystallization of the austenite, in other words, the precipitates form both in the grain boundaries and inside the austenitic crystals favoring the stability of the non-recrystallized microstructure. After the allotropic transformation takes place, the microstructure is formed by very fine ferrite grains due to the fact that this phase nucleates both in the grain boundaries as well as inside the crystals forming subgrains, deformation bands or structural defects (Figure 1.5).

Inside the crystal

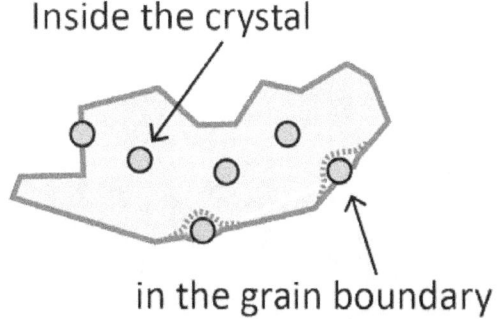

in the grain boundary

Figure 1.5 Austenite crystal deformed with carbides and nitrides precipitates both inside the crystal and at the grain boundary

1.2.2 Titanium Nitrides

If the starting temperature for the precipitation of titanium nitrides is 1355°C (Table 1.1), Figure 1.6 shows the graphic evolution of the titanium and nitrogen contents at different temperatures: at 1628 K (precipitation temperature) and two other inferior ones (1373 K and 1273 K). This figure shows the amount of N and Ti dissolved in the material, in other words, it has not precipitated yet. The line obtained by linear regression represents the evolution of the N and Ti content on this example steel as a function of precipitation. Furthermore, almost all nitrogen precipitates forming TiN at 1373 K (1100°C); at the delay time, the temperatures changes from 1075°C to 829°C, and almost all the N has already precipitated and little is left behind to form other precipitates.

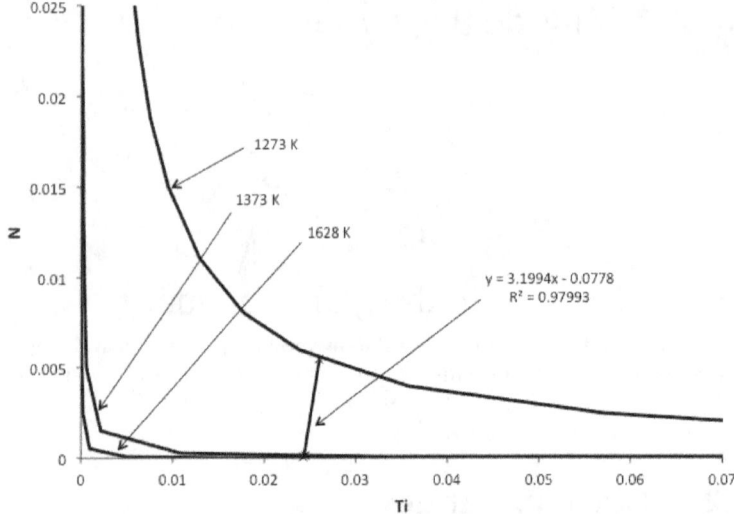

Figure 1.6 Evolution of titanium and nitrogen contents at different temperatures

1.2.3 Niobium carbides

If the starting temperature for the precipitation of niobium carbides is 1157°C (Table 1.1), Figure 1.7 shows the graphic evolution of the niobium and carbon content at different temperatures: at 1430 K (precipitation temperature) and two other inferior ones (1373 K and 1073 K). It also shows the amount of Nb and C dissolved in the example material, in other words, it has not precipitated yet. The line obtained by linear regression represents the evolution of the C and Nb content of this steel as a function of precipitation. Moreover, at a temperature of 1073 K, relatively all the Nb has precipitated during the delay time (1075~829°C); this explains the role of this microalloying element: inhibit the grain growth during the waiting time and strongly delaying the static recrystallization of the austenite in the finishing stage, stimulating the fine ferrite grain size after the allotropic transformation and a slight structural hardening as a result of the nanometric size of precipitates.

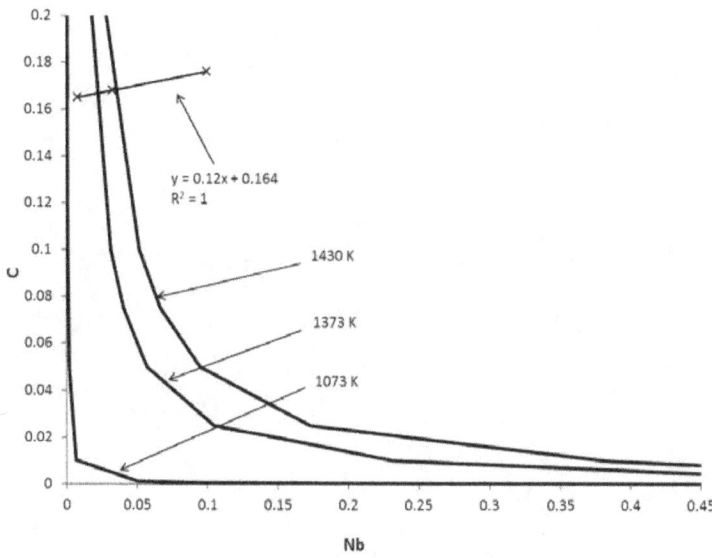

**Figure 1.7 Evolution of niobium and carbon contents
at different temperatures**

If SEM microscopy is used in order to analyze different types of carbides, Figure 1.8 shows a titanium carbonitride, identified by the shape of almost a rectangle; also, element mapping confirms that niobium precipitates form using the titanium as a base because the titanium carbonitrides have formed at higher temperatures. Figure 1.9 presents a series of titanium carbonitrides together with niobium carbides (grown over the TiCN) that precipitated on a pearlitic band.

Figure 1.10 shows another type of precipitates, in this case a vanadium carbide formed in the grain boundary of a ferrite grain, while Figure 1.11 shows a titanium carbonitrade that acted a substrate to the growth of not only a niobium carbide but also a vanadium carbide. Figure 1.12 shows two carbides: a vanadium carbide and a very small molybdenum carbide.

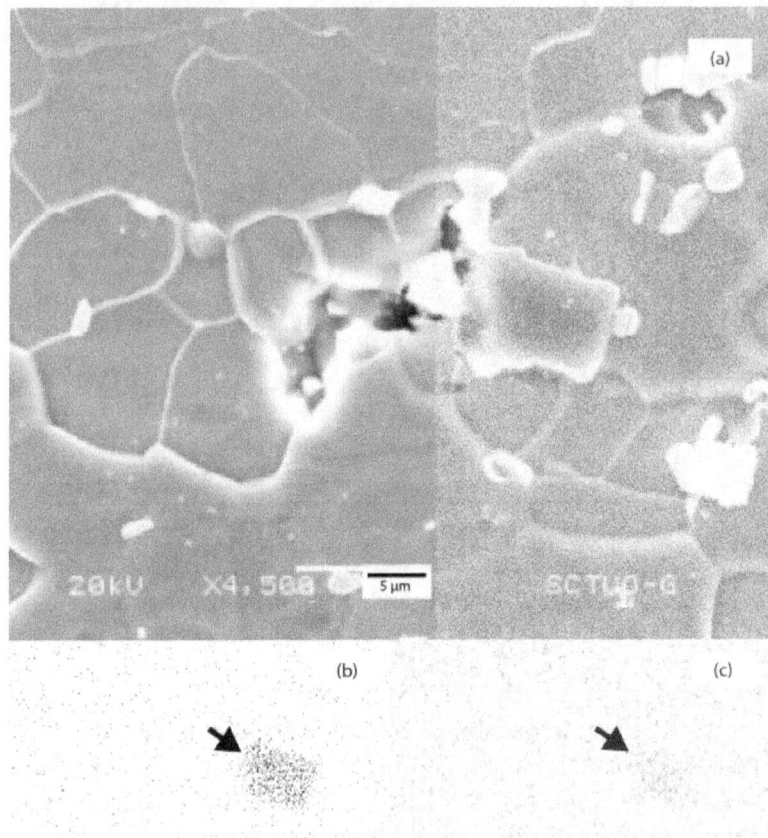

Figure 1.8 SEM micrograph showing a titanium carbonitride and niobium carbide (a), dot mapping of titanium (b) and dot mapping of niobium (c)

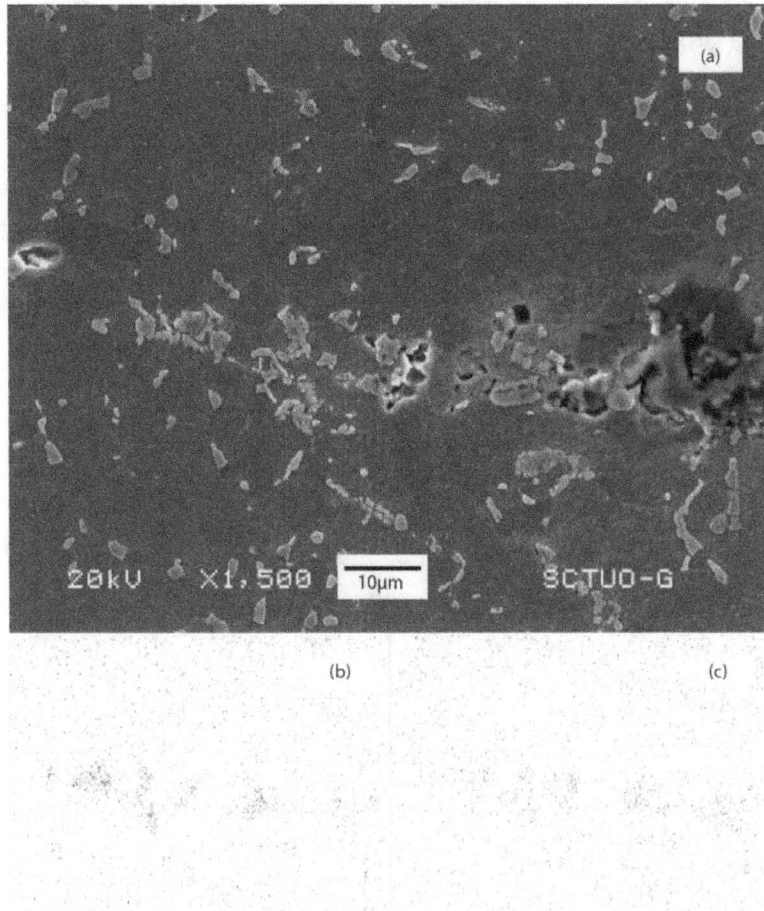

Figure 1.9 SEM micrograph showing a string of titanium carbonitrides
and niobium carbides (a), dot mapping of titanium (b) and
dot mapping of niobium (c)

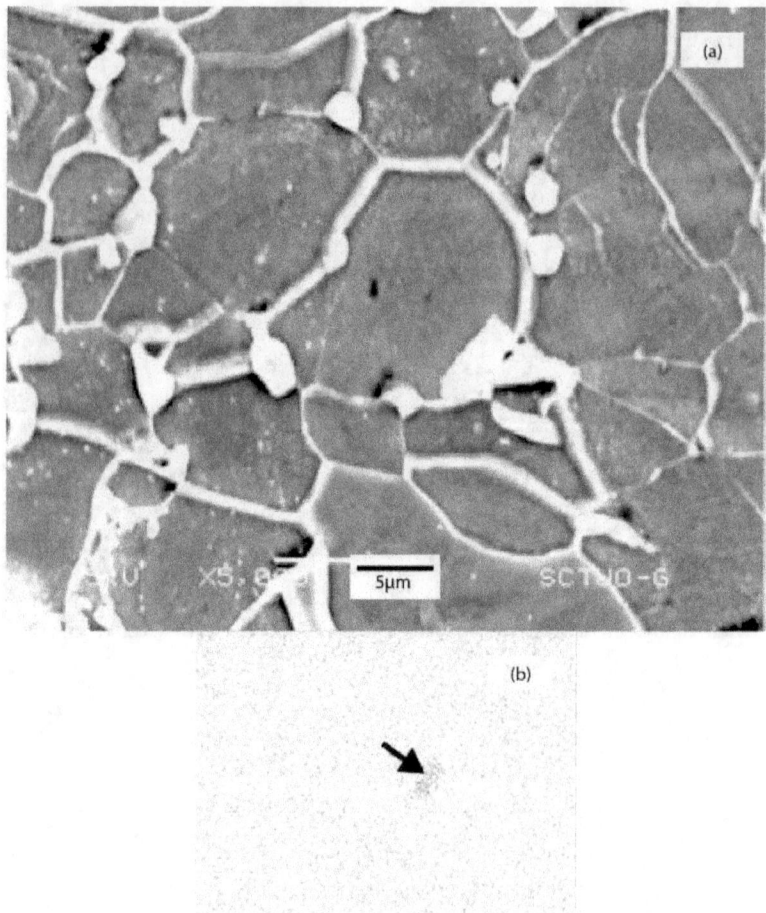

Figure 1.10 SEM micrograph showing a vanadium carbide (a) and dot mapping of vanadium (b)

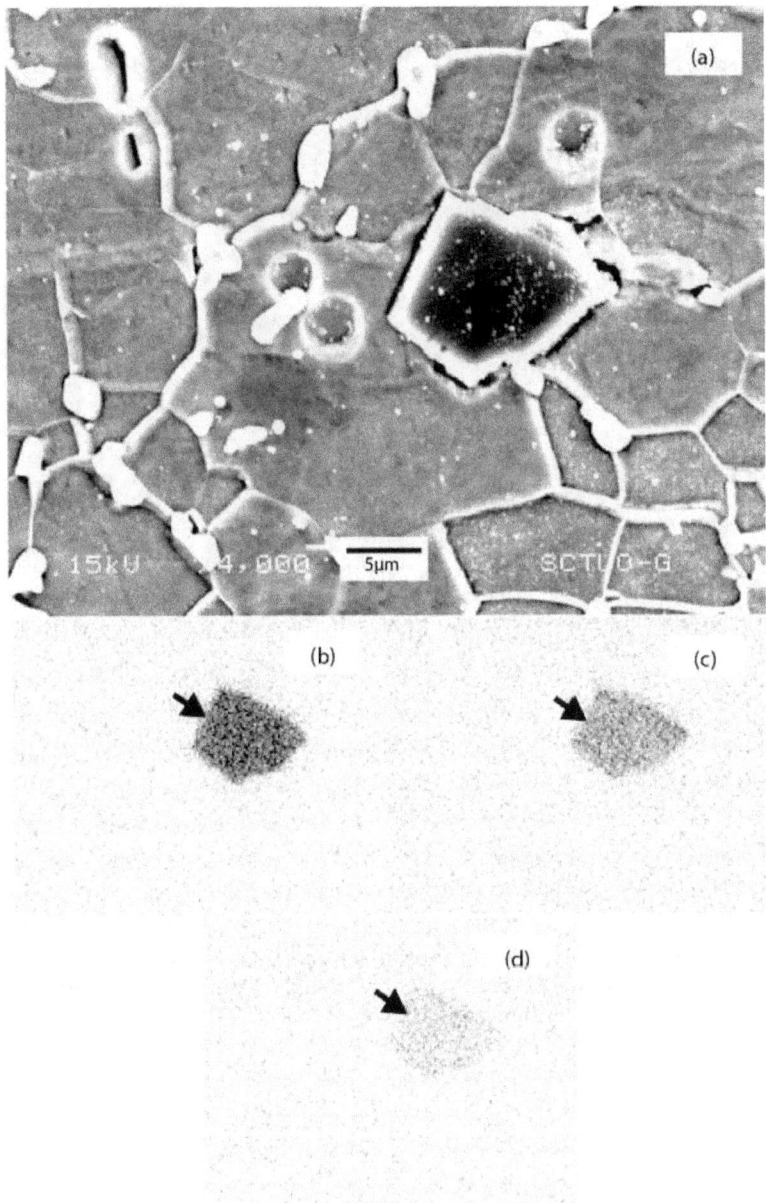

Figure 1.11 SEM micrograph showing a titanium carbonitride,
a niobium carbide and vanadium carbide (a), dot mapping
of titanium (b), dot mapping of niobium (c) and
dot mapping of vanadium (d)

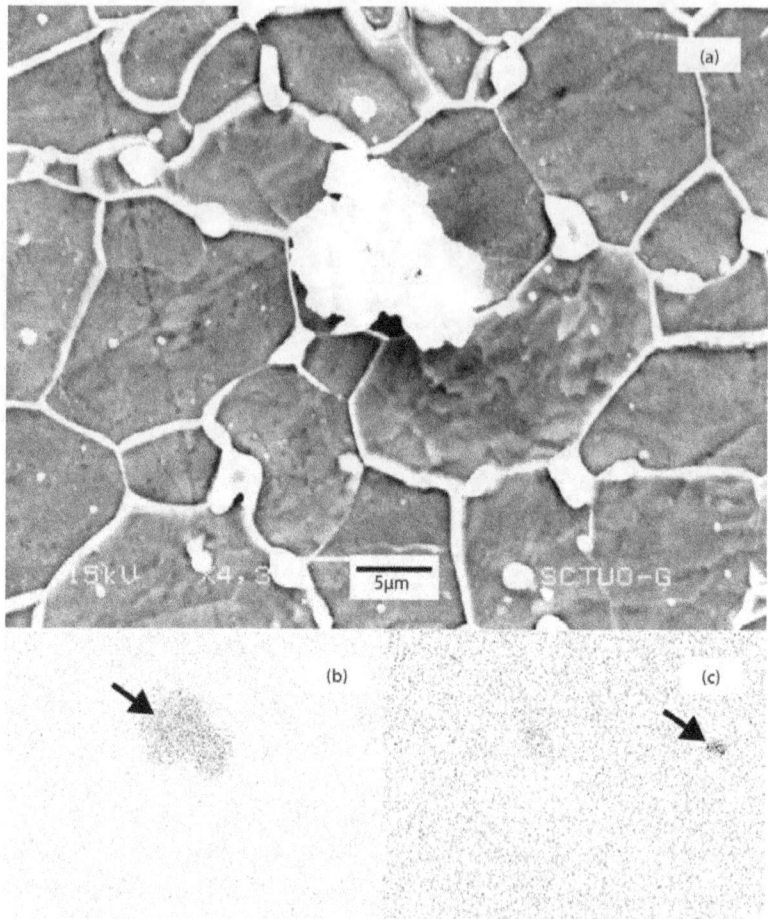

Figure 1.12 SEM micrograph showing a vanadium carbide and
a molybdenum carbide (a), dot mapping of vanadium (b) and
dot mapping of molybdenum (c)

The variety of geometries, sizes and morphologies of precipitates observed by SEM in these types of steels, are product of the combination of different alloying elements capable of forming, along with carbon and nitrogen dissolved in the steel, second-phase particles which prevent the growth of grain size during the different stages of the rolling process. In this manner, Ti is a key allowing element as its capacity to precipitate at high temperatures and control austenite grain size, combines with its role as substrate for Nb and V, resulting in ferrite-pearlite banded microstructures with both large and small precipitates located either at the grain boundaries or inside the grains. Only a combination of such different parameters may allow quality control processes where the mechanical strength of the steel is not sacrificed in order to allow formability and industrial manufacture of parts.

1.3 References

1. *The Quiet Revolution in Materials Manufacturing and Production.* **Eager, T.W.** 4, 1998, Journal of Metals, Vol. 50, p. 19.
2. *Advances in the Hot Rolling of Steels.* 1, 2004, Ironmaking & Steelmaking, Vol. 31, p. 8.
3. **Pero-Sanz, J.A.** *Steels: Physical Metallurgy, Selection and Design.* Madrid : CIE-Dossat, 2000.
4. *The Changing Scene in Steel.* **Paxton, H.W.** 12, 1979, Metallurgical Transactions, Vol. 10A, p. 1815.
5. **Pickering, F.B.** *Physical Metallurgy and the Design of Steels.* s.l. : Applied Science Publishers, 1978.
6. *A study on the microstructural changes in hot rolling of dual-phase steels.* **Salehi, A.R., Serajzadeh, S. and Karimi, A.** 2006, Journal of Materials Science, Vol. 41, pp. 1917-1925.
7. *Static Strain Aging Phenomena in Cold-Rolled Dual-Phase Steels.* **Waterschoot, T., De, A.K. and Vandeputte, S.** March 2003, Metallurgical and Materials Transactions A, Vol. 34A, pp. 781-791.
8. *Dual-Phase Ultrafine Grained Steels Produced by Controlled Rolling Processes.* **Quintana, M.J., Gonzalez, R. and Verdeja, L.F.** [ed.] MS&T. Columbus, Ohio : s.n., 2011. MS&T 2011 Proceedings.

9. *Ultrafine Grained HSLA Steels for Cold Forming.* **González, R., García, J.O. and Barbés, M.A.** 10, 2010, Journal of Iron and Steel Research, International, Vol. 17, pp. 50-56.

10. *Strength and ductility of ultrafine grained aluminum and iron produced by ARB and annealing.* **Tsuji, N., Ito, I. and Saito, Y.** 2002, Scripta Materialia, Vol. 47, pp. 893-899.

11. *Ultrafine grained structure formation in steels using dynamic strain induced transformation processing.* **Beladi, H., Kelly, G.L. and Hodgson, P.D.** 2007, International Materials Review, Vol. 52, pp. 14-28.

12. *A new route to fabricate ultrafine-grained structures in carbon steels without severe plastic deformation.* **Okitsu, Y., Takata, N. and Tsuji, Y.** 2009, Scripta Materialia, Vol. 60, pp. 76-79.

13. *Rapid Transformation Annealing: A novel method for grain refinement of cold rolled low-carbon steels.* **Lesch, C., Alvarez, P. and Bleck, W.** 9, 2007, Metallurgical and Materials Transactions, Vol. 38A, p. 1882.

14. *Deformation Characteristics Evaluation of Modified Equal Channel Angular Pressing Processes.* **Yoon, S.C., Nagasekhar, A.V. and Yoo, J.H.** 1, 2010, Materials Transactions, Vol. 51, pp. 46-50.

15. *Influence of High-Pressure Torsion Straining Conditions on Microstructure Evolution in Commercial Purity Aluminum.* **Todaka, Y., Umemoto, M. and Yamazaki, A.** 1, 2008, Materials Transactions, Vol. 49, pp. 7-14.

16. *Technical parameters affecting grain refinement by high pressure torsion.* **Hohenwarter, A., Bachmaier, A. and Gludovatz, B.** 12, 2009, International Journal of Materials Research, Vol. 100, pp. 1653-1661.

17. *Numerical Cooling Strategy Design for Hot Rolled Dual Phase Steel.* **Suwanpinij, P., Togobytska, N. and Prahl, U.** 11, 2010, Steel Research International, Vol. 81, pp. 1001-1009.

18. **Pero-Sanz Elors, J.A.** *Ciencia e ingeniería de materiales: Estructura, transformaciones, propiedades y selección.* [ed.] CIE Dossat. 5th. 2006. (in spanish).

19. *Controlled Rolling of Steel Plate and Strip.* **Tanaka, T.** 1981, International Metals Reviews, Vol. 4, pp. 185-191.

20. *Influence of Thermo-Mechanical Processing Parameters and Chemical Composition on Bake Hardening Ability of Hot*

Rolled Martensitic Steels. **Asadi, M. and Palkowski, H.** 7, 2009, Steel Research International, Vol. 80, pp. 499-506.

21. *Effect of Composition and Process Variables on Nb (C,N) Precipitation in Niobium Microalloyed Austenite.* **Dutta, B. and Sellars, C.M.** 3, 1987, Materials Science and Technology, Vol. 3, p. 197.

22. *Modelling the Kinetics of Strain Induced Precipitation in Nb Microalloyed Steels.* **Dutta, B., Palmiere, E.J. and Sellars, C.M.** 5, 2001, Acta Materialia, Vol. 49, p. 785.

23. *Nucleation Kinetics of Ti Carbonitride in Microalloyed Austenite.* **Liu, W.J. and Jonas, J.J.** 4, 1989, Metallurgical Transactions, Vol. 20A, p. 689.

24. *Deformation Processing.* **Backofen, W.A.** 12, 1973, Metallurgical Transactions, Vol. 4B, p. 2679.

25. **Porter, D.A., Easterling, K.E. and Sherif, M.Y.** Phase Transformations in Metals and Alloys. 2009, 2, pp. 63-110.

26. **Pero-Sanz, J.A.** *Ciencia e Ingeniería de Materiales.* Madrid : CIE-Dossat 2000, 2006.

2
Room-temperature behavior

DP steels provide a good example of an UFG material ($d \leq 5$ μm) and are suitable for the production of automotive parts that require high mechanical resistance, high impact strength, and high elongation. They are produced using low-alloy steels as a basis, reducing costs and resulting in a combination of martensite and ferrite structures with ultrafine grain sizes. The desirable characteristics are obtained through strict control of rolling conditions, namely, strain rate, cooling rate, and direct quenching. The room-temperature mechanical behavior of DP steels, along with their microstructural characterization, is presented to clarify the effect of the ATMCRP used to produce these materials.

2.1 Mechanical behavior

Construction steels (materials used in civil engineering) are cheap and weldable and have good mechanical resistance and fracture toughness. Their microstructures are usually formed by ferrite and pearlite, and sometimes include bainite, martensite, room-temperature retained austenite (the base of the TRIP steels), and even nanoprecipitates of Ti, Nb, and V. Although the main phase of these steels is ferrite, certain construction steels (DP steels) present two phases: a soft phase such as ferrite and a hard phase, which is commonly martensite with traces of bainite, usually in the form of a dispersion. The excellent mechanical properties of these steels allow the production of mechanical components with thinner sections than those manufactured from conventional steels. These mechanical properties include a high strength/elongation ratio, excellent impact response, soft yield behavior, low yield-stress/tension-stress ratio, and high formability. The strength of these steels is related to the amount of plastic deformation applied during the thermomechanical processes in the intercritical region, and results from the formation of substructures in the ferrite. Epitaxial ferrite (the phase that grows during thermomechanical processing) is responsible for the improvement in tension stress as the transverse area decreases during rolling. These improvements in mechanical properties are achieved without significant reduction in the ductility of the steel[1; 2; 3].

The ferrite grain size d is the sole parameter responsible for simultaneously increasing the mechanical resistance (yield stress σ_y) and increasing the toughness (i.e., decreasing the temperature of the ductile–fragile transition, *ITT*) of a steel; moreover, grain refinement of a steel increases the value of the yield stress/*ITT* ratio, optimizing thickness and cost, and making the material structurally and economically more efficient. These three variables (grain size, yield stress, and *ITT*) can be related as follows[4; 5; 6; 7]:

$$\sigma_y \approx 54 + 17d^{-1/2}$$

[2.1]

$$ITT \approx -19 - 11.5\,d^{-1/2} + 2.2\,(\text{pearlite \%})$$

[2.2]

where d is expressed in mm, σ_y in MPa, and *ITT* in °C.

Thus, a ferritic–pearlitic steel with a grain size ASTM G of 10 ($d \approx 10\,\mu m$) and 15% of pearlite (approximately 0.1% C) would have $\sigma_y \approx 225$ MPa and $ITT \approx -100°C$. These ferritic grain sizes (10 μm) are usually obtained by a normalizing thermal treatment after hot forging of the steel. Ultrafine ferritic steels (UFF) have grain sizes lower than 5 μm (12 ASTM G) and are usually manufactured by ATMCRP in a continuous annealing processing line (CAPL).

Figure 2.1 shows the stress–strain engineering curves for DP600 and DP780 steels tested at room temperature; it is evident that DP780 exhibits a higher yield stress than DP600, with a tension stress close to 800 MPa, while that for DP600 is close to 650 MPa. Also, DP600 has a higher elongation (%A) and well-defined yield stress (threshold activation stress for the dislocation movement), in contrast to DP780, where the elastic limit is not evident from the curve and must be graphically obtained (e.g. parallel line at 0.2%).

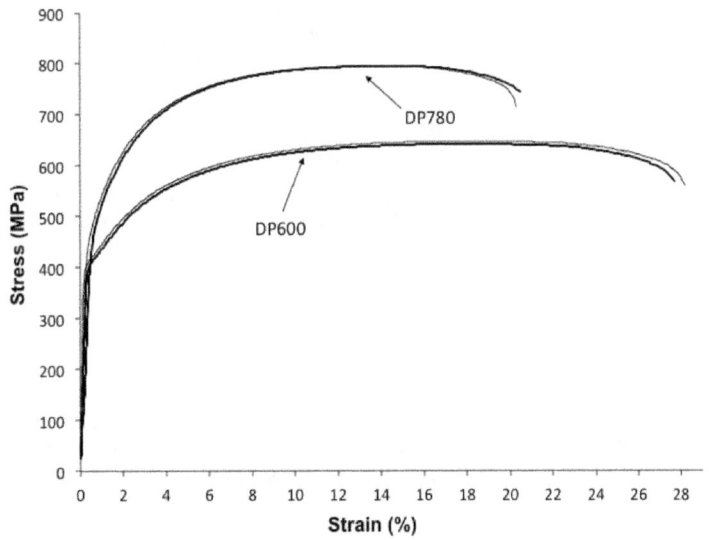

Figure 2.1 Stress–strain engineering curves for 2 DP steels tested at room temperature

The linear regression curve of true yield versus true maximum stress (logarithms) for plastic deformation can be obtained (Figure 2.2) in order to calculate the strain-hardening coefficient n from the equation:

$$\log \sigma = n \log \varepsilon + \log K$$

[2.3]

where n is the slope of the straight line obtained by linear regression. In both cases, these values are higher than 0.15, in other words high enough for manufacturing processes of bending or drawing, although DP780 has a higher value owing to its higher strength and lower plasticity.

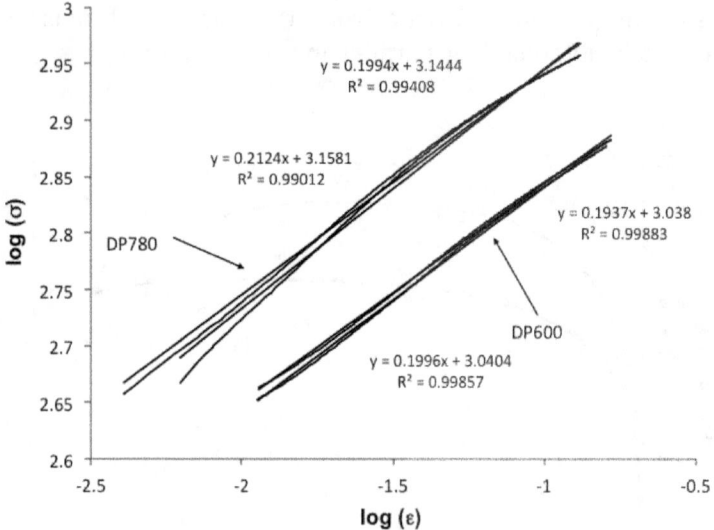

Figure 2.2 Linear regression for the strain-hardening coefficient n for DP600 and DP780 steels, including the regression equation and correlation coefficient R^2

2.2 Microstructure

Figure 2.3 and Figure 2.4 show the results of quantitative metallography for DP600 and DP780 steels, respectively: the microstructure (a), identification of the ferrite and martensite phases by a Buehler Omnimet apparatus connected to a microscope (b), the ferrite grain size histogram (c), and an evaluation of amount of the martensitic grains (d). It can be seen from Figure 2.3a that the DP600 steel has a slightly equiaxed microstructure as well as some grains elongated in the rolling direction (horizontal); this effect is evident in the banded structure of the DP780 steel. The distribution of the ferrite grain size indicates that in DP600 (Figure 2.3c) this can be approximated by a normal distribution with mean grain size between 14 and 15 ASTM G; this is in contrast to what happens in DP780, which has a combination of large and small grains, with a mean size between 17 and 18 ASTM G.

Thanks to the microstructure (phases present, amount of each phase and its distribution) induced by the ATMCRP, the products made with these steels have mechanical properties such as yield stresses up to 600 MPa and tension stresses close to 800 MPa, which are only possible for a low alloy steel if the formed microstructure is of the ultrafine grained type ($>$ 14 ASTM G, $<$ 2.5 μm) along with precipitates of Ti, Nb and V.

On the other hand, it is also very important that the strain hardening coefficient n is higher than 0.15 to avoid instabilities and defects during the drawing and/or bending processes used in the manufacture of products from the automotive industry among others. Otherwise, the steel may have very limited or non-practical applications, and the data obtained for similar microstructures produced at laboratory scale is not relevant for its industrial uses.

The compromise between an UFG microstructure (grain sizes close to or smaller than 1 μm) and high values of n, is possible because of the internal stresses suffered by the ferrite caused by the transformation of austenite into martensite during cooling.

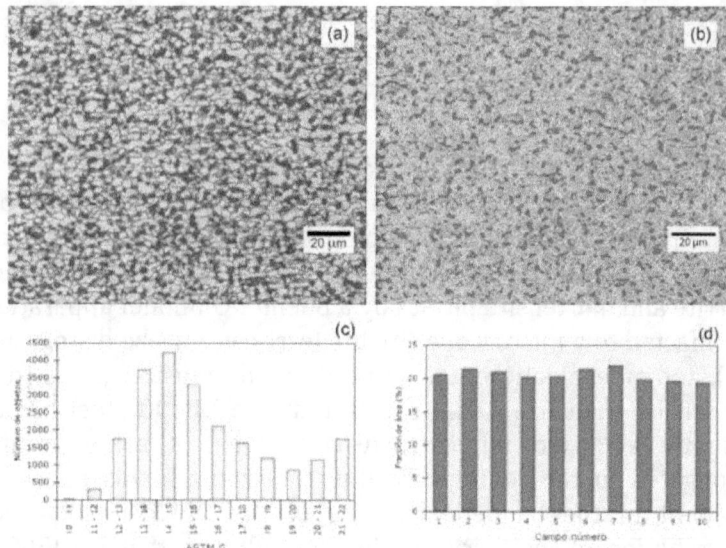

Figure 2.3 Quantitative metallography of DP600 steel: (a) microstructure; (b) phase identification; (c) ferrite grain size histogram; (d) percentage of martensite

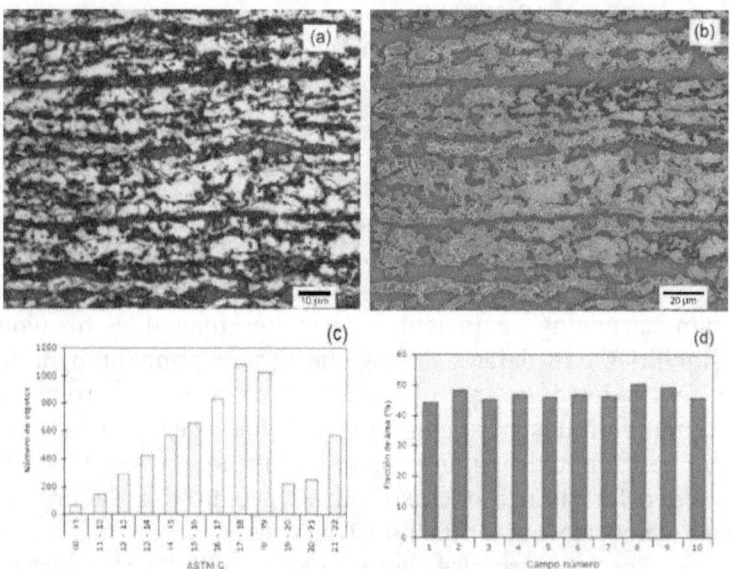

Figure 2.4 Quantitative metallography of DP780 steel: (a) microstructure; (b) phase identification; (c) ferrite grain size histogram; (d) percentage of martensite

2.3 References

1. *Grain refinement in Dual-Phase Steels.* **Mukherjee, K., Hazra, S.S. and Militzer, M.** September 2009, Metallurgical and Materials Transactions A, Vol. 40A, pp. 2145-2159.
2. *Effect of Martensite Plasticity on the Deformation Behavior of a Low-Carbon Dual-Phase Steel.* **Mazinani, M. and Poole, W.J.** February 2007, Metallurgical and Materials Transactions A, Vol. 38A, pp. 328-339.
3. *Effect of rolling and epitaxial ferrite on the tensile properties of low alloy steel.* **Ahmad, E., Sarwar, M. and Manzoor, T.** 2006, Journal of Materials Science, Vol. 41, pp. 5417-5423.
4. **Pickering, F.B.** *Physical Metallurgy and the Design of Steels.* s.l. : Applied Science Publishers, 1978.
5. **Pero-Sanz Elors, J.A.** *Ciencia e ingeniería de materiales: Estructura, transformaciones, propiedades y selección.* [ed.] CIE Dossat. 5th. 2006. (in spanish).
6. *The Deformation and Ageing of Mild Steel: III. Discussion of Results.* **Hall, E.O.** 1951, Physical Society of London Proceeding, Vol. 64, pp. 747-753.
7. *The Cleavage Strength of Polycrystals.* **Petch, N.J.** 1953, Iron and Steel Institute Journal, Vol. 174, pp. 25-28.

3
High-temperature behavior

HSLA steels with ultrafine grains can exhibit superplasticity when traction-tested at specific temperatures and with low strain rates ($\dot{\varepsilon} < 10^{-2}\,\text{s}^{-1}$), showing very high elongations (>100%) without localized necking and with mainly intergranular fractures. This behavior requires the initial grain size to be small (<10 μm) so that the flow of mass can be inhomogeneous (sliding and rotation of the grain boundaries, accommodated by diffusion). After superplastic deformation, a steel of this type may exhibit decohesion of the hard (pearlite and carbide) and ductile (ferrite) phases. Comparisons of experimental data and mathematical models for different grain sizes indicate that the Ashby–Verrall model provides an explanation of the intergranular sliding of the ferrite/ferrite or ferrite/pearlite grain boundaries, with dislocation creep (power-law creep) and diffusion creep (linear-viscous creep) mechanisms occurring simultaneously.

3.1 Superplasticity

A material can be considered superplastic when, owing to the strong interdependence of the creep tension stress and the strain rate, necking is either absent or very slight (a series of diffuse necks) along a part or test zone of a specimen. In other words, the material elongates uniformly when traction-tested, reaching elongations greater than 100% and in some cases 1000% (Figure 3.1). That is, superplasticity can be described as the ability of crystalline solids (metals) to achieve extremely high uniform elongations when tension-tested: for these materials, strength is highly sensitive to strain rate. This phenomenon surpasses plasticity, and therefore the presence of dislocations (linear defects that slide over crystallographic planes in lattice directions) provokes plastic and irreversible deformations in each crystal or grain of the material[1; 2; 3; 4].

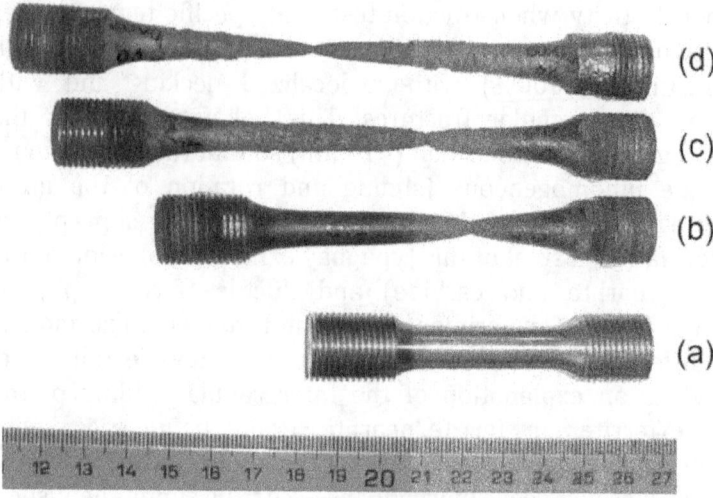

Figure 3.1 Elongation of samples with original $L_0 = 30$ mm (a) after traction testing at 800°C with crosshead speeds of 0.5 (b), 0.2 (c) and 0.1 (d) mm/min

From a microstructural point of view, superplasticity is achieved when two phenomena take place simultaneously in the material, namely, grain boundary migration and grain boundary shearing/sliding. Both theoretical models and microstructural evidence indicate that the most important aspect of this behavior is grain boundary sliding (GBS); nevertheless, the presence of dislocations or diffusion in grains or in zones near grain boundaries is necessary in order to maintain the superplasticity of the material[4; 5].

In order to understand superplasticity, it is necessary to analyze creep (i.e. flow of material under constant load and at a constant temperature, which has to be higher than 0.3 T_M). A typical creep curve (Figure 3.2) has three stages: primary transitory creep (parabolic), secondary stationary creep (linear), and tertiary exponential creep (where cavitations occur). In the second stage, the strain rate is constant and is related to the load (stress) applied and the diffusional phenomena (mass transport), depending on temperature according to an Arrhenius-type law[6].

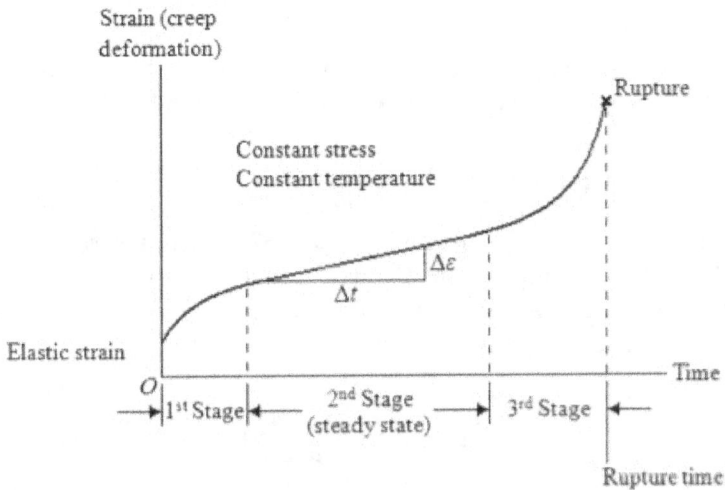

Figure 3.2 Typical creep curve

There are two laws determining the behavior of creep by sliding (conservative movement) and climbing (non-conservative movement) of dislocations[7; 8; 9]:

- *Power-law creep, or Weertman creep*[10]:

$$\dot{\varepsilon} = A\sigma^n \exp\left(-\frac{Q}{RT}\right)$$

[3.1]

where A and n are constants, σ is the stress, Q is the activation energy (equal to self-diffusion) of creep, R is the ideal gas constant, and T is the absolute temperature.

- *Exponential-law creep, or Dorn creep*[11]:

$$\dot{\varepsilon} = A\sinh(\beta\sigma) \exp\left(-\frac{Q}{RT}\right)$$

[3.2]

where A and β are constants and the remaining variables are the same as in equation [1.3].

Both types of behavior require high stresses, but these are always lower than the yield stress, at intermediate $(0.3T_M < T < 0.5T_M)$ or high $(T > 0.5T_M)$ temperatures. At intermediate temperatures, mass transport may take place through processes such as grain boundary diffusion, with a diffusion coefficient D_{it} given by:

$$D_{it} \approx D_v \left(1 + \frac{\delta}{d}\frac{D_{gb}}{D_v}\right)$$

[3.3]

where D_v is the volume diffusion coefficient, D_{gb} is the grain boundary diffusion coefficient, δ is the width of the grain boundary (which is approximately equal to two interatomic distances, $2b$), and d is the grain size. At low temperatures, the controlling process is grain boundary diffusion, while volume diffusion is more important at higher or intermediate temperatures[7; 8; 9].

The diffusion creep mechanism occurs at low applied stresses (lower than those required for sliding or dislocation climbing) compared with other types of diffusion, but at similar intermediate or high temperatures, as previously mentioned. Under the effect of applied stress, grains are deformed either by intergranular diffusion (at intermediate temperatures) or by volume diffusion (at high temperatures), without the intervention of dislocations. Moreover, there is immediate generation of cavities or voids, which can be related to the sliding of the grain boundaries that is generated by the diffusion processes and is necessary to maintain, at least temporarily, the continuity of the material (mass). The law determining this behavior (Herring–Nabarro creep) is of the Newtonian or viscous-flow type[12; 13]:

$$\dot\varepsilon = \frac{B\sigma \exp(-Q/RT)}{d^2}$$

[3.4]

where B is a constant, d is the grain size, and the remaining variables are the same as in equation [3.1].

In order to obtain superplasticity, a fine-grained, equiaxed, and stable microstructure is required (close to 1 μm) in order to produce intergranular deformation and dislocation climbing, although the smaller grain size results in a greater number of grain boundaries, thus promoting boundary sliding and reducing the distance for accommodation by diffusion and/or slip[1; 2; 6; 14; 15; 16; 17].

As indicated by Wadsworth et al. [4], "The fine-structured steel exhibits ideal characteristics in that it is weak and superplastic (~1000% tensile elongation) at warm temperatures and strong and ductile at room temperature, in contrast to the coarse-structured steel which is strong and not very ductile at warm temperatures and relatively weak at room temperature".

A disadvantage of the excessively fine-grained structure is the decreasing ductility due to the negative influence on the strain-hardening coefficient n (from the expression $\sigma = K\varepsilon^n$), which may block the uniform elongations required in cold-forming operations such as folding or biaxial expansion of low-carbon steel. The high-yield-stress and weldable steels required for industrial applications may exhibit inhomogeneous and unstable deformations (the Piobert–Lüders effect), introducing superficial defects and premature necks in formed products[18].

If the relation between stress and strain rate is expressed as $\sigma = K\dot{\varepsilon}^m$, another requirement is that the coefficient m must have values between 0.3 and 0.7, the latter being the maximum reached experimentally, which means that it is at least one order of magnitude higher than the value for hot-deformation processes and similar to the values for forming polymers and glasses. The yield stresses are dependent to a small degree on the strains, but are very sensitive to the strain rate and temperature, as shown in Figure 3.3[19].

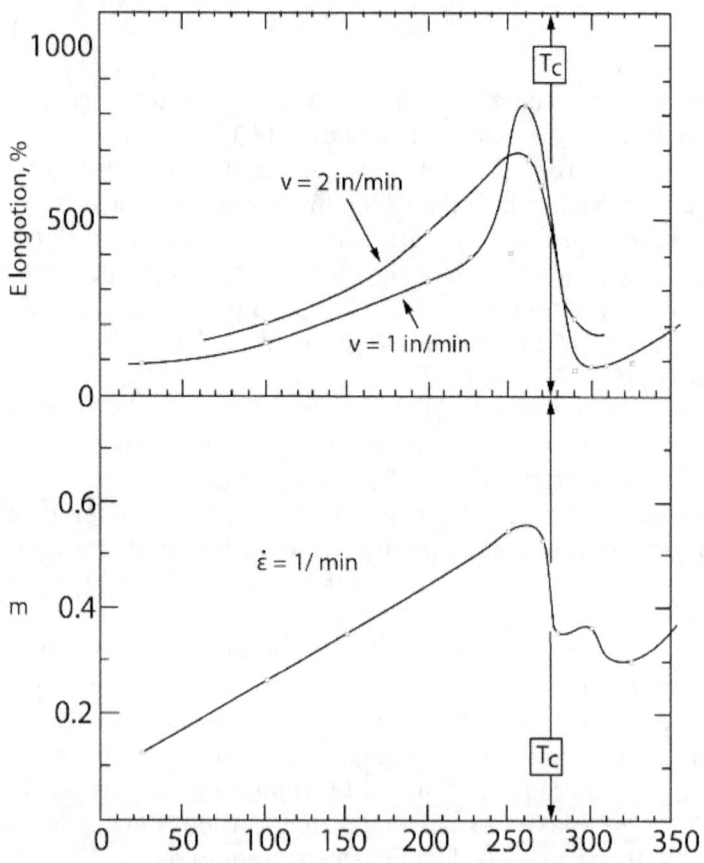

Figure 3.3 Temperature dependence of elongation (upper) and strain-rate sensitivity (lower). T$_C$ is the critical temperature[19]

Furthermore, superplasticity requires forming temperatures higher than $0.5\,T_M$ (i.e. 50% of the melting temperature), since this process is diffusion-controlled and deformation of the material must occur under conditions where the diffusive flow is reasonably fast. At these temperatures, the grain size must remain stable and without growth, and corrosion must be prevented by using an inert atmosphere or a protective varnish. After superplastic forming of the material, it will not behave superplastically at temperatures lower than $0.2\,T_M$. Its room-temperature mechanical characteristics will depend on its nature and the ultrafine grain size, but can include high values of yield stress, tension stress, hardness, toughness, and fatigue stress, among others[1; 2; 6; 16].

Additionally, slow strain rates $(10^{-3}\sim10^{-1}s^{-1})$ are required, but these are impossible to obtain with conventional forming techniques such as rolling or extruding, and when the necessary techniques are available, they are frequently unsuitable for a production environment. Furthermore, the stress required to deform the material has to be sensitive to the strain rate such that if necking starts, with the affected area deforming much more rapidly, the resulting increase in strain rate will harden the material at this point until necking stops and there is a return to uniform deformation. A high strain-rate sensitivity provides stability against localized necking and thus allows high plastic elongation. However, the strain rate is related to grain size in an inversely proportional way as:

$$\dot{\varepsilon} \approx (\mathbf{b}/d)^p$$

[3.5]

where $\dot{\varepsilon}$ is the strain rate, $\mathbf{b}$ is the Burgers vector, d is the grain size, and $p = 2$ when the process is related to lattice-diffusion-controlled creep (Nabarro–Herring creep) or $p = 3$ when it is related to grain-boundary diffusion-controlled creep (Coble creep). In other words, if the grain size is decreased, a higher strain rate can be used to achieve the same state of superplasticity [1; 2; 6; 20].

Finally, the grain boundaries of the material have to allow grain sliding between them and grain rotation when stress is applied, owing to the movement of dislocations along the grain boundaries. In order for this to happen, it is necessary to apply the right temperature and to have a fine-grained structure. These requirements are fundamental in some superplastic behavior models, such as that proposed by Ashby and Verrall. The sliding percentage in the superplastic region is very high (40~80%), which makes a considerable contribution to the maximum superplastic deformation, although this is the case only for elongations <50%; for elongations greater than this, the major contribution is from diffusion. The sliding percentage decreases considerably at both low and high strain rates, as does the overall elongation[6; 15; 21].

On the other hand, Alden[22] lists nine microstructural characteristics required of a material for it to deform superplastically (some of which have already been mentioned):

1. A strong dependence of stress on strain rate:

$$\sigma = K\dot{\varepsilon}^m$$

[3.6]

2. An influence of temperature on the diffusional processes related to superplasticity. The activation energies are similar to those for grain boundary diffusion and not to those for volume diffusion.
3. Creep at a constant load and temperature that does not exhibit a primary or transitory phase.
4. A material that, after being superplastically deformed, retains its initial resistance and ductility at room temperature.
5. A superplastic material that, after deformation, does not exhibit substructures or an increase in dislocation density.
6. A crystal shape factor (length/width) that is low and unrelated to superplastic deformation. The number of grains in the transverse section of a sample should decrease on deformation.
7. The absence of texture related to large superplastic deformations. The grains should not tend to

reorient themselves (crystallographically speaking) in relation to the main material flow directions.

8. Superplasticity associated with intergranular sliding and rotation of grains during deformation, meaning that they constantly switch their relative positions while accommodating themselves in the mass flow direction.

9. Stable and fine-grained microstructure.

It is generally agreed that when a material with equiaxed microstructure is deformed under optimum superplastic conditions, the shape of the crystals does not change and they grow very little, although it should be remembered that GBS creates precipitate-free zones near grain boundaries[3].

The dual structure of the material is important with regard to superplastic behavior, since the second phase (30~50% in volume) can inhibit grain growth in the matrix by stabilizing the microstructure. Therefore, single-phase materials cannot reach superplasticity by themselves[2; 16; 17; 23].

3.2 Mechanical behavior

If an HSLA steel is tensile tested at different temperatures between 600°C and 900°C at 50°C intervals and different crosshead speeds in order to determine the temperature interval at which the steel would present superplasticity, Figure 3.4 shows the engineering traction curves for samples tested at 5 mm/min. As expected, the higher the temperature, the lower the maximum stress the material can withstand; for 600°C this is above 200 MPa, while for 900°C it is below 80 MPa. It is noteworthy that at 800°C the elongation of the material is greater than 100%. At temperatures above 750°C and below 850°C, the result is smooth deformation and very high elongation, indicating superplasticity.

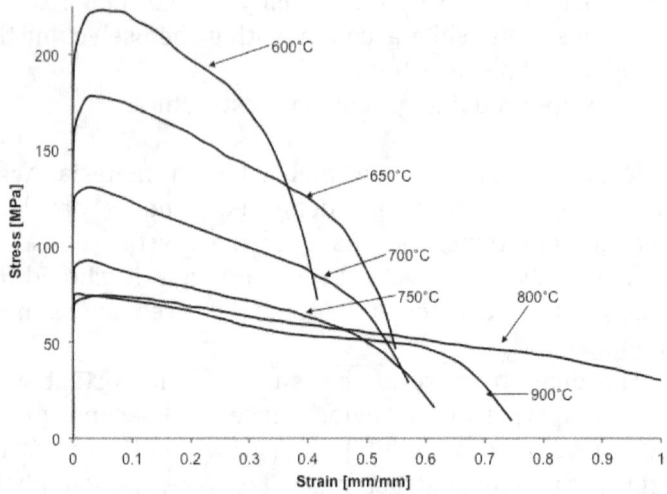

Figure 3.4 Engineering stress–strain curves at different temperatures and 5 mm/min crosshead speed

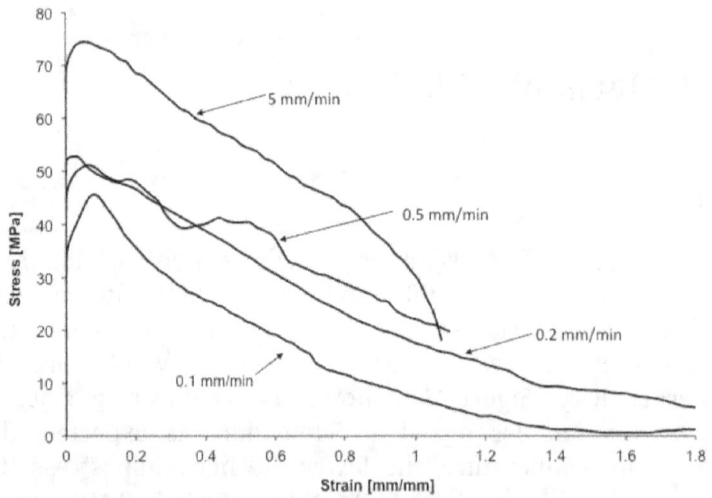

Figure 3.5 Engineering stress–strain curves at 800°C and different crosshead speeds

The results of tests at 800°C are shown in Figure 3.5, where, although some ripples are evident during deformation at 5 mm/min, many more ripples are observed at 0.5 mm/min. Smooth deformation of the samples was only achieved when the crosshead speed was lowered to 0.2 mm/min. For speeds of 0.2 and 0.1 mm/min, the elongation of the samples was close to 200%, also presenting diffuse necking.

Table 3.1 Tension testing results obtained with different strain rates at 800°C

L_0	CS	$\dot{\varepsilon}$	S_y	elong
30		2.83×10^{-5}	25 (creep)	-
30	0.05	2.78×10^{-5}	27.4	137.5
30	0.1	5.56×10^{-5}	34.4	181.7
57	0.2	5.85×10^{-5}	37.6	>110.0
30	0.2	1.11×10^{-4}	52.80	191.3
57	0.5	1.46×10^{-4}	45.8	>110.0
30	0.5	2.78×10^{-4}	57.3	92.7
57	5	1.46×10^{-3}	70	>110.0
57	10	2.92×10^{-3}	82.4	84.2
30	10	5.56×10^{-3}	86.2	105
23	20	1.45×10^{-2}	93.68	126.6

L_0 - specimen initial length (mm)
CS - Crosshead speed (mm/min)
$\dot{\varepsilon}$ - Strain rate (s^{-1})
S_y - Yield stress (MPa)
elong - elongation (%)

Table 3.1 shows the values of the yield stress (σ_y) and strain rate ($\dot{\varepsilon}$) obtained from tests at 800°C. Furthermore, according to expression [3.6] ($\sigma = K\dot{\varepsilon}^m$), K is a function of the temperature, the previous deformation the steel may have suffered and the grain size; coefficient m expresses the sensibility of the applied tension to the strain rate as follows:

$$m = \left(\frac{\log(\sigma_{y_2}/\sigma_{y_1})}{\log(\dot{\varepsilon}_{0_2}/\dot{\varepsilon}_{0_1})} \right)_{T,d,\varepsilon}$$

[3.7]

where σ_y is the yield stress at 0.2% and $\dot{\varepsilon}_0$ the initial strain rate, in tests made at two different strain rates. Thusly the dependence of stress and strain rate is shown in Figure 3.6a where a clear zone II behavior can be seen, also, the transition from zone I (creep) and zone II is not evident as much lower strain rates should be tested in order to observe it. Moreover, at higher strain rates, zone III emerges. The regression lines presented in the figure have slope values of ~0.6 for zone II and ~0.1 for zone III.

Figure 3.6b shows the m coefficient from equation [3.7] as a function of strain rate using pairs of data obtained from Table 3.1. When this coefficient has values between 0.3 and 0.7, superplastic behavior is achieved[24], in this case the maximum value of m will be reached at a strain rate close to 2.8×10^{-5} s^{-1}. In other words, at 800°C this will be the best deformation rate in order to obtain superplasticity in this steel.

The behavior of the steel tested at high temperatures (Figure 3.5 and Figure 3.6b) presents three strain mechanisms related to the crosshead speed (time elapsed in the test):

1. A first one associated to the deformed ferrite, dynamically restored (5 mm/min curve).
2. A second one initially superplastic, but with non-stable microstructure, continued by grain growth that induces hardening (ferrite restoration) and concluded by localized necking (conventional) (0.5 mm/min curve).
3. A third one, also superplastic, that applies to stable or nearly stable structures that do not stop the intergranular straining associated to superplasticity. The mechanisms of straining by sliding and climbing of dislocations, as well as intergranular sliding accommodated by diffusion mainly at the grain boundaries (Ashby-Verrall model) are the dominating ones (0.2 mm/min curve).

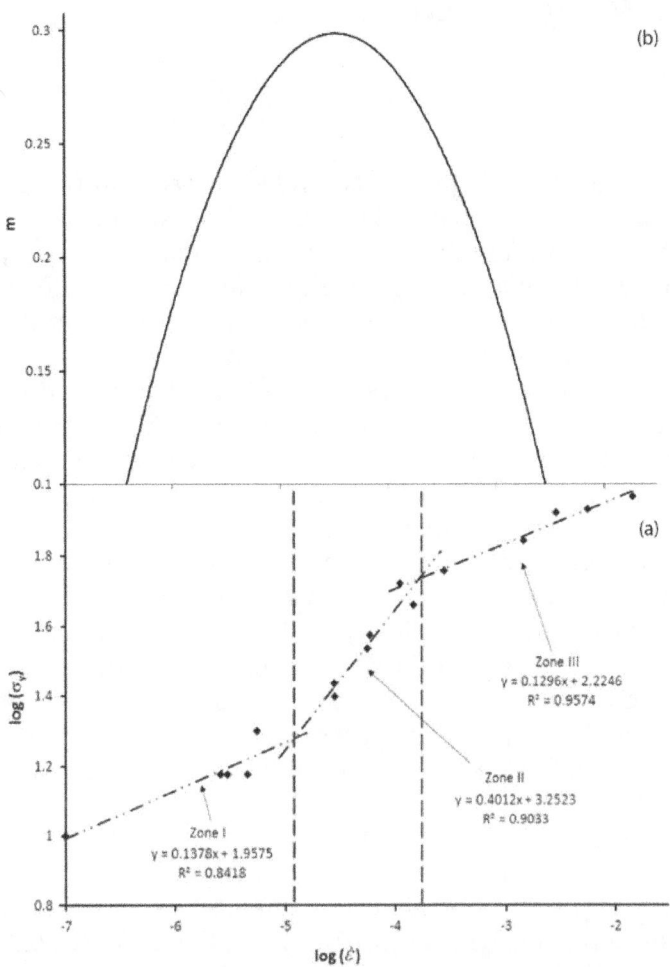

**Figure 3.6 Influence of strain rate on yield stress (a)
and super-index m (b) of equations [3.6] and [3.7]
in superplastic behavior at 800°C**

3.3 Microstructure

Superplastic materials can be divided into two groups depending on the type of fracture they exhibit: necks with substantial cross-sectional area or necks shaped as a very fine point when fractured. They all exhibit internal cavitations after fracture, but not all show voids (as voids form, their distribution depends on the strain rate). Higher strain rates produce small voids that are arranged like chains in a direction almost parallel to the deformation one, while lower strain rates produce larger voids that are mostly round or spherical and are apparently distributed in a random manner[25; 26].

An explanation for this behavior is that, with high strain rates, a large number of void nuclei start to form with a chainlike distribution, and, for sufficiently short test periods, the voids grow only slightly and do not reach the critical radius. This difference in void morphology is also consistent with the transition of the deformation mechanism of the material from power-law creep at high strain rates to grain boundary diffusion (or vacancy diffusion) at low strain rates. The behavior of the critical void radius, which increases with decreasing strain rate, is also consistent with the transition between mechanisms[25; 26; 27; 28; 29]. Langdon[25] also indicates that sometimes voids interlink themselves, usually in the tensile direction, which is probably due to the high ductility of the material (superplasticity).

Other authors have established that cavities nucleate and grow at the grain boundaries when undergoing high-temperature creep, and also when the material is behaving superplastically (especially UFG materials). The reason for this behavior is that, owing to the ultrafine nature of the grains, the voids at the boundaries sometimes involve several grains, which enhances the diffusion process; this enhancement is even greater when coalescence of voids also occurs. It is believed that GBS is responsible for the nucleation of voids and cavities in superplasticity, but this requires the grains to have high angles of disorientation, and therefore the processing of the material to obtain ultrafine grain sizes has a direct influence on the possibility of reaching superplastic behavior. In other words, the more disordered the boundary structure, the greater the ability of GBS to produce UFG material will be, because of the high density of broken bonds across the boundary; the dual structure in steels suitable for GBS can be formed during the early stage of deformation[2; 14; 15; 16; 17; 25; 26].

A grain boundary may be of the non-equilibrium type when there are some defects in addition to the equilibrium content, for example non-equilibrium grain boundaries produced by the absorption of lattice dislocations. These high-energy defects are unstable and disappear at high temperatures, but they increase both the number of vacancies at boundaries and the diffusion coefficient. At these grain boundaries, certain kinetic processes related to diffusion (such as migration and GBS) are accelerated, which in turn affects creep, recrystallization, and superplasticity, since these result from the interaction of grain boundaries with lattice dislocations (movement of a grain boundary through a strained crystal or when a dislocation enters a grain boundary during plastic deformation). In other words, superplastic flow occurs through GBS as the individual grains of the polycrystalline matrix move over each other in response to applied stress (Rachinger sliding), and this happens without any elongation of the individual grains. Non-equilibrium grain boundaries are commonly found in ultrafine- and hyperfine-grained materials[16; 30; 31].

Figure 3.7 shows a representation of GBS in superplasticity: dislocations move along the grain boundary between two grains and accumulate at the junction of three grains (A); this concentrates the stress, and the slip nucleates in the contiguous grain; the dislocations move across the grain, pile up at the opposing grain boundary (B), and are then removed by climbing into the boundary. Also, the accommodating dislocation can glide through the blocking grain to produce effects on the opposing grain boundary[16]. According to Valiev et al.[30], "the degree of the non-equilibrium grain boundary structure is determined by the straining parameters: temperature and strain rate."

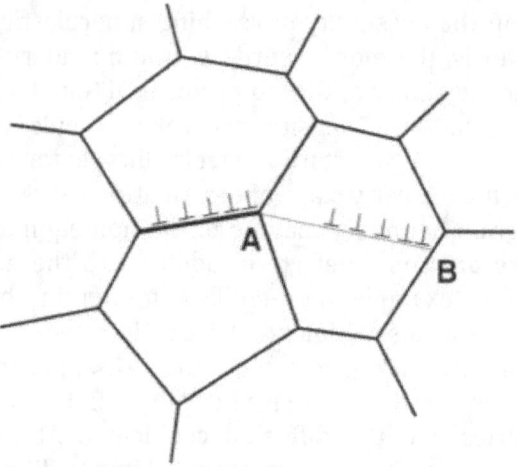

Figure 3.7 Principles of a model for grain boundary sliding in superplasticity: dislocations move along the grain boundary and pile up at the triple junction A; the stress concentration is removed by nucleation of slip in the adjacent grain, and these intergranular dislocations pile up at B and climb into the grain boundary [16]

If the raw hot rolled state microstructure of an HSLA steel is compared to another one tested at 800°C and 0.1 mm/min (Figure 3.8a and b), it is clear that the bands have not completely disappeared though the ferrite phase has suffered restoration without a significant grain size enlargement. If these two microstructures are then compared to a steel tested at 750°C and 0.1 mm/min (Figure 3.8c), a clear difference can be observed, as testing at a lower temperature results in a complete dissapearence of the pearlitic bands, being replaced by ferrite grains and precipitates (mostly carbides).

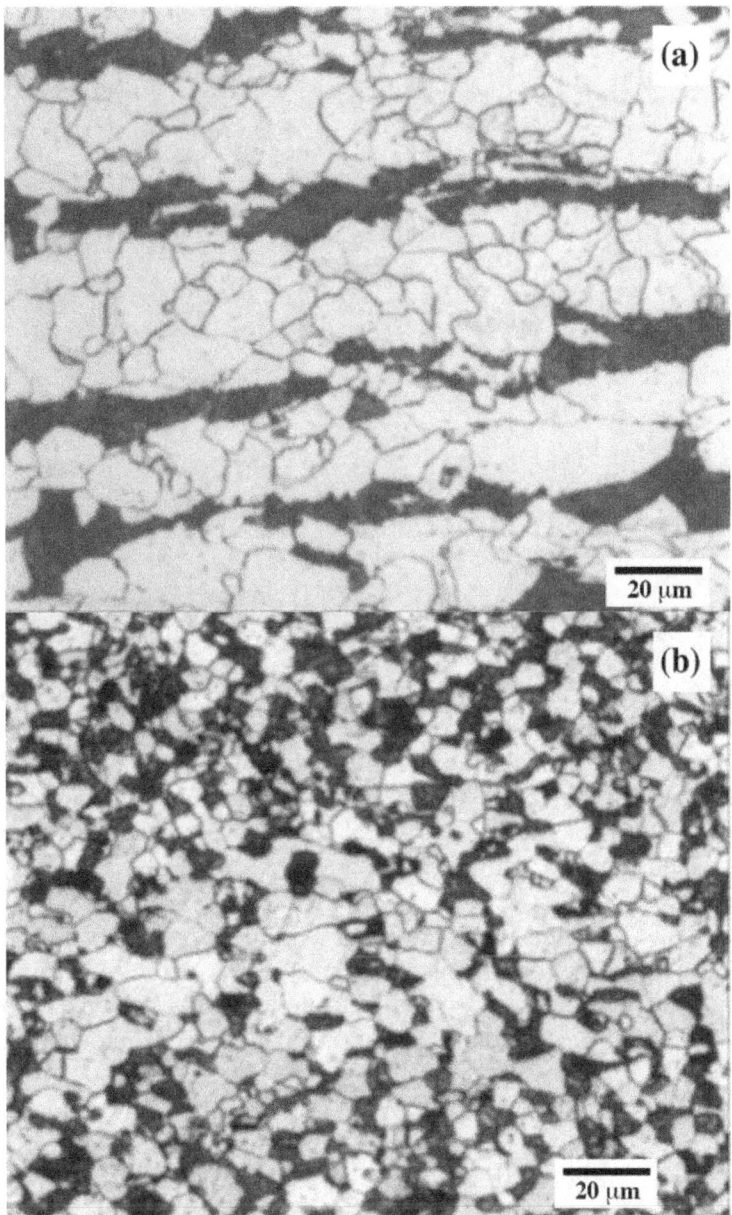

Figure 3.8 Microstructure of a steel in its raw hot rolled state (a), tested at 800°C and 0.1 mm/min crosshead speed (b) and tested at 750°C and 0.1 mm/min crosshead speed (c)

cont. Figure 3.8

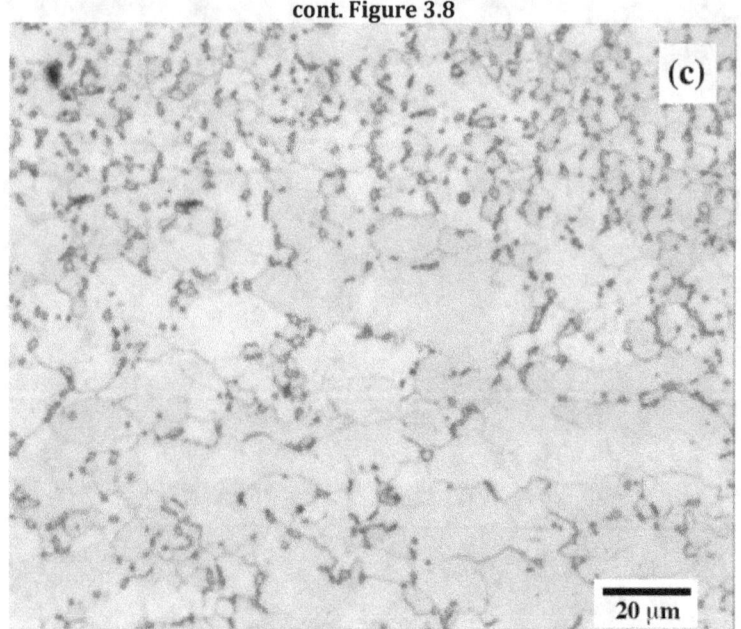

Figure 3.9 presents the microstructure after superplastic deformation: ferrite grains of a larger size which are slightly elongated in the rolling direction with evidence (subgrains) of having suffered dynamic recovery during deformation, and also,

- decohesions shaped as w and r, mainly located in the ferrite/pearlite (previous austenite grains) interphase, which is an unequivocal proof of intergranular sliding during the deformation process
- small cavities in the α-pearlite (previous austenite grains) interphase, which shows the different deformation capacity for each of these phases
- null evidence of generalized grain growth during deformation, as grain size is very similar (or even slightly smaller, <5 μm) to the raw hot rolled state one
- grain (or grain clusters) sliding and rotating, as a consequence of superplastic deformation

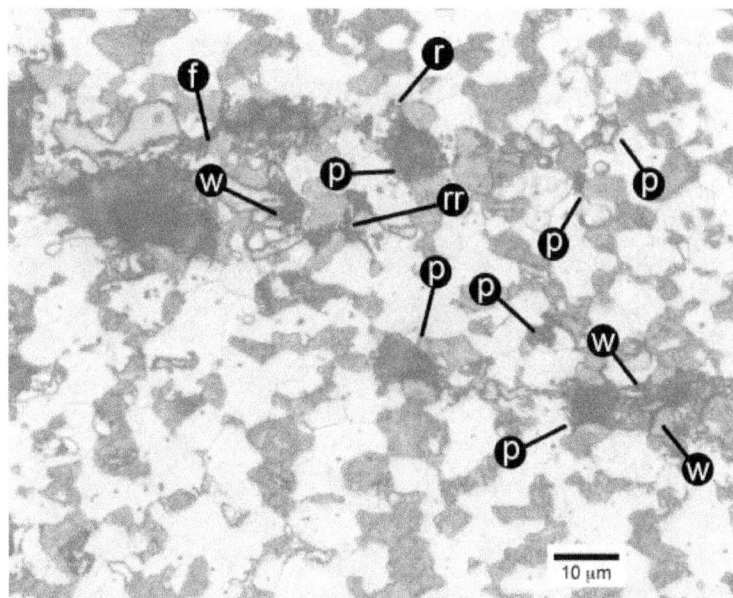

Figure 3.9 Micrograph of a specimen superplastically deformed at 800°C at a zone close to rupture (15 mm away from it). w-shaped decohesion between ferrite-ferrite-pearlite (w), rr-shaped decohesion between ferrite and pearlite (rr), ferrite-pearlite decohesion (f) and ferrite-ferrite (r) or pearlite-pearlite decohesion (p) are observed.

On scanning electron micrographs of the superplastically deformed steel, w-shaped (Figure 3.10) as well as double r-shaped (Figure 3.11) decohesions can be seen. These indicate detachment between the matrix (ferrite) and pearlite (precipitates), leading to a ductile fracture (Figure 3.12), very pronounced and practically in the shape of a pencil point. Intergranular straining of the ferrite (soft/superplastic constituent) and its intrusion into the austenite (hard constituent), maintaining the grain size of the steel, appear to be necessary conditions for deformation of the steel sheet. Higher proportions of ferrite (lower straining temperatures), by limiting the proportion of austenite and increasing its instability, completely eliminate the superplastic behavior.

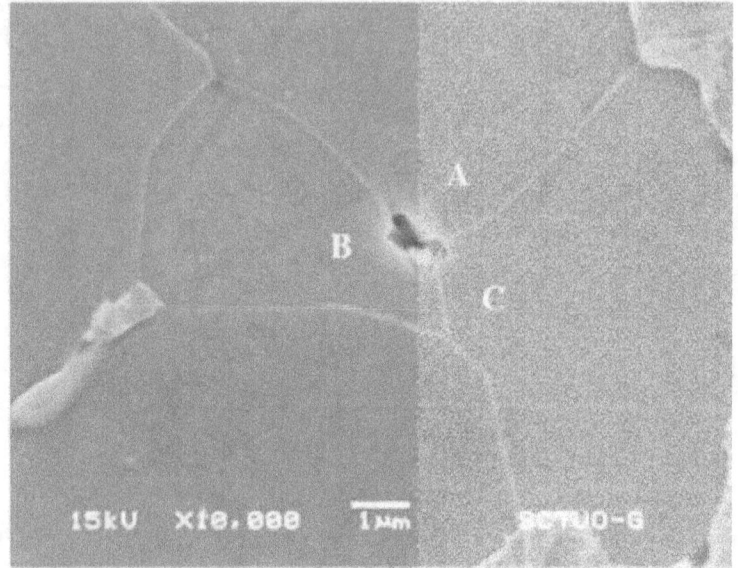

o *w-shaped* cracks

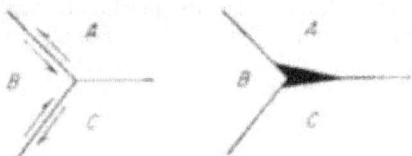

**Figure 3.10 SEM micrograph of the steel tested at 800°C
showing a w-shaped decohesion**

Figure 3.11 shows evidence of decohesions which are evidence of superplastic behavior in the steel and consequence of the grain sliding. Also, Figure 3.10 shows another type of decohesion resulting from boundary sliding of three grains (one grain displaces the other two).

Furthermore, decohesions between ferrite and pearlite may result in cavities with evidence of ductile deformation of the softer phase, as shown in Figure 3.12. Superplastic flow stops when the intergranular damage and decohesions between the matrix and inclusions leads to ductile fracture.

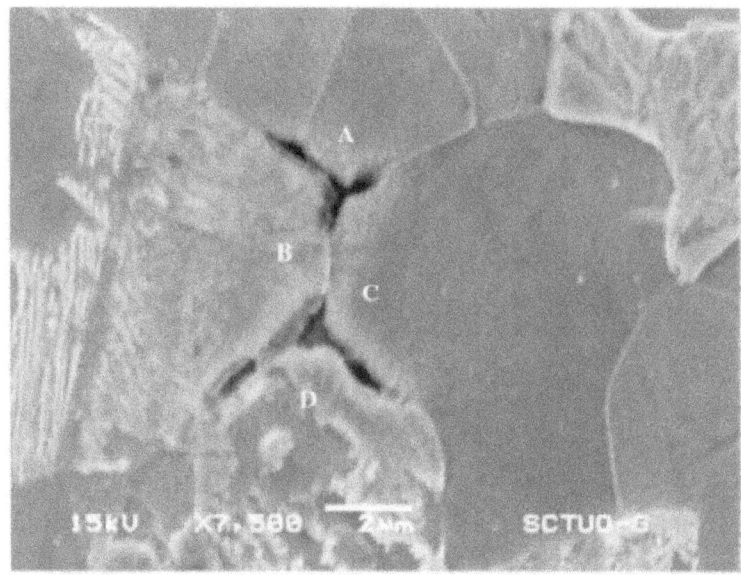

○ *r-r-shaped* cracks

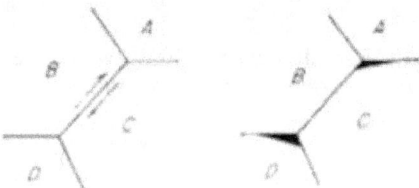

Figure 3.11 SEM micrograph of the steel tested at 800°C
showing a r-r-shaped decohesion

The importance of decohesions is evident if the role of Ti and Nb carbides (or carbonitrides) is taken into account as these particles are not dissolved during the 800°C test (Figure 3.12): these precipitates anchor the grain boundary and prevent recrystallization and grain growth during rolling at the finishing stage, which is accompanied by the formation of $\gamma -$ pancaked grains and deformation bands. Consequently, a larger number of nucleation sites are made available for the $\gamma \rightarrow \alpha$ transformation. This allows the formation of an ultrafine microestructure, fulfilling the requirements for both strength and toughness[32; 33].

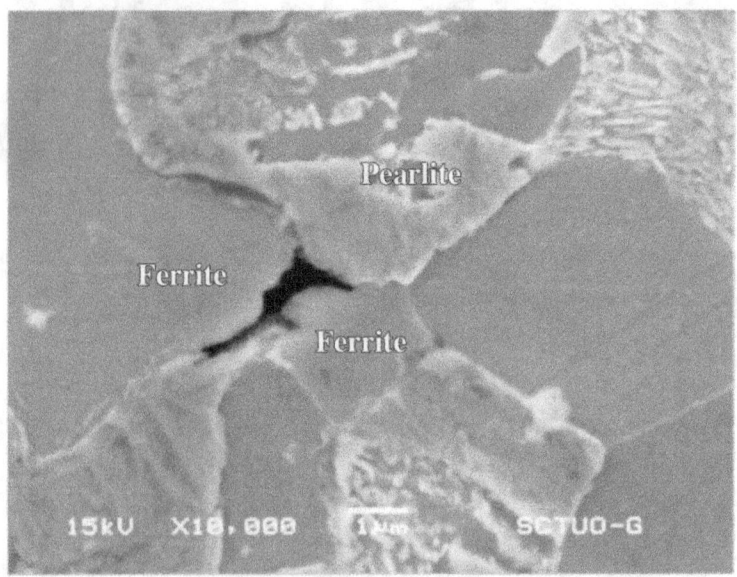

Figure 3.12 SEM micrograph of the steel tested at 800°C showing a ferrite-pearlite (ductile) decohesion

The initial composition (Nb and Ti content) of the steel, and the evidence of an ultrafine microstructure after ATMCRP indicate that the role of NbC and Ti(C,N) are very important during both room temperature and high temperature deformation of the material. Figure 1.11 shows a particle precipitated at grain boundaries which was initially formed as a titanium carbonitride and then became a substrate for a niobium carbide layer, as confirmed by other authors[32].

If at a lower magnification, precipitates are observed (Figure 1.9), it is evident that most of the titanium carbonitrides grow along the former pearlite bands, and as previously mentioned the niobium carbides appear at the same spots using the titanium precipitates as a substrate.

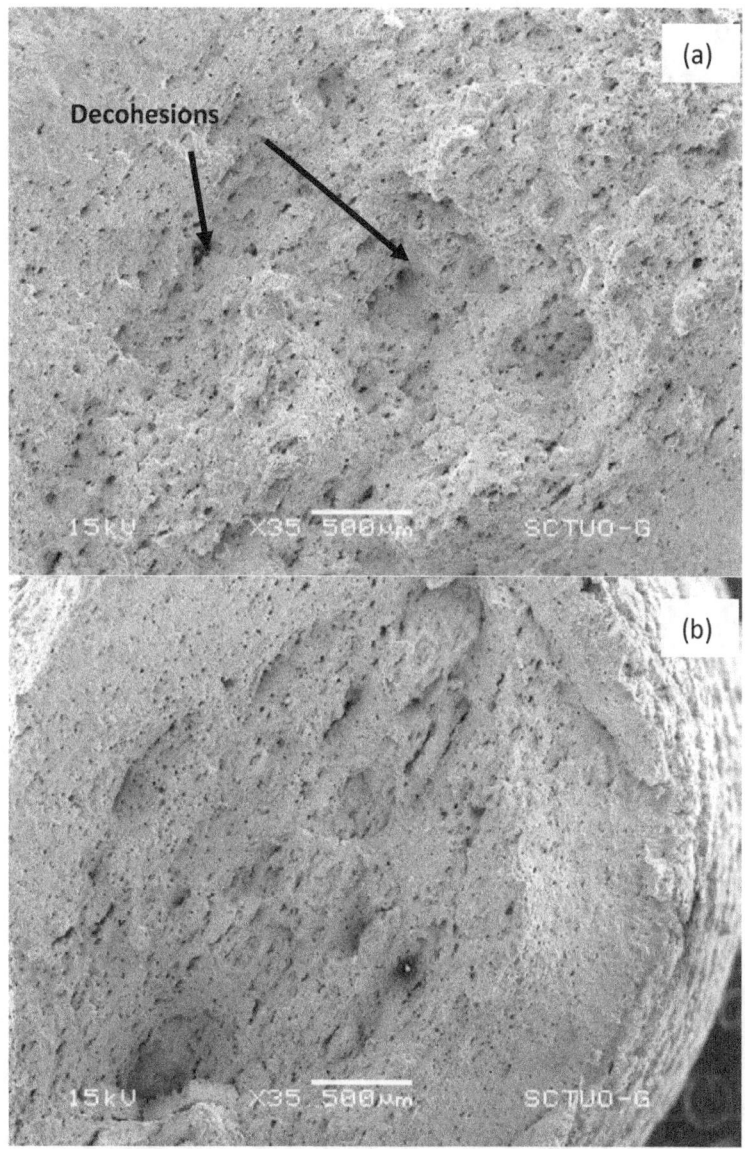

Figure 3.13 SEM fracture surface micrographs of steels tested in tension in their raw hot rolled state (a) and superplastically deformed and then room temperature tested (b)

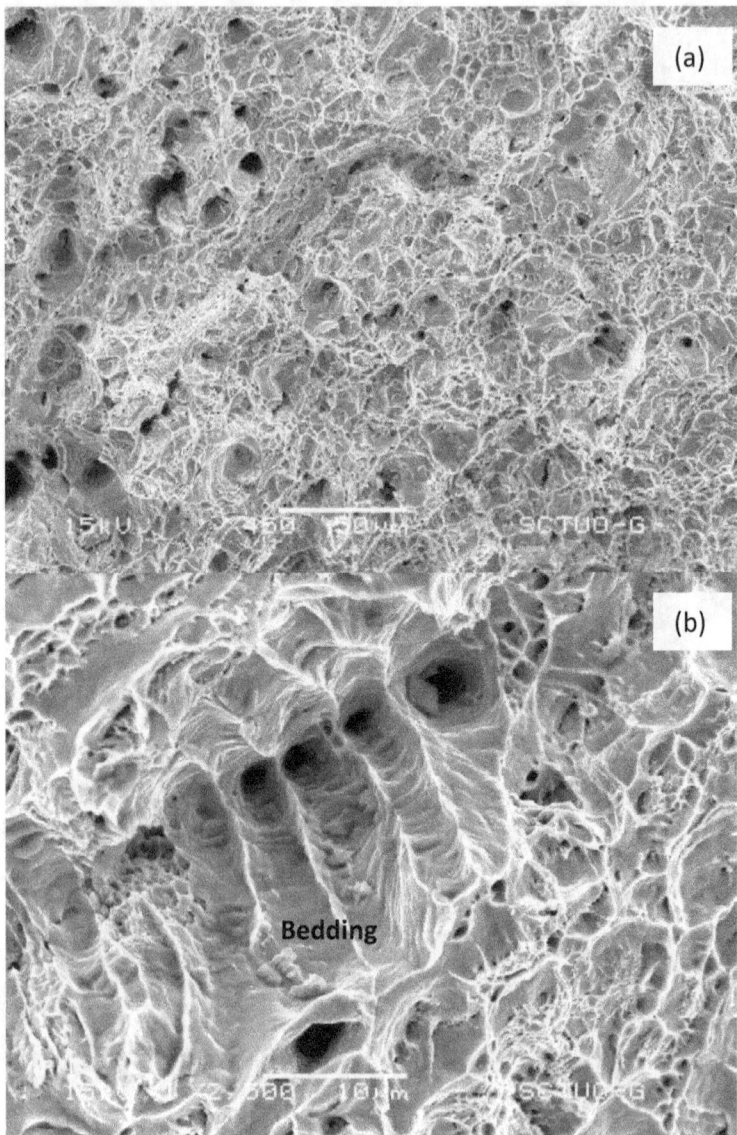

Figure 3.14 SEM micrographs of the fracture surface of a steel superplastically deformed then room temperature tension tested

Figure 3.13 shows the SEM micrographs of the fracture surface of both a steel tested in its raw hot rolled state and one superplastically deformed and then room temperature tension tested. In both cases decohesions are orientated with the main axis of the ellipse formed at the neck of the fracture surface. The superplastically deformed material shows a smaller amount of decohesions though with larger size. Nevertheless, both fracture behaviors appear to be very similar.

High magnification SEM micrographs of the superplastically deformed fracture surface are shown in Figure 3.14, where the following may be noticed: large and small dimples (a), caused mainly by the decohesion of MnS and matrix (large ones) and possibly caused by the decohesion of (Ti,Nb) (C,N) and matrix (small ones).

The difference in size of the decohesions observed on the fracture surface of the specimens (Figure 3.13) indicates that the superplastically deformed steel forms larger decohesions as these last ones were previously formed during the superplastic test at 800°C.

If the mechanisms of fracture of the superplastically deformed material are considered, the presence of dimples and bedding (Figure 3.14) shows an enhanced capacity of the ferrite phase to deform while precipitates must suffer decohesion from the matrix in order to allow fracture. Figure 3.14b shows in its upper zone a precipitate with an almost squared geometric shape along with decohesions (dark zones) left by pearlite or precipitates when pulled out of the matrix.

3.1.1 Subgrains

Besides precipitation and defects caused by these last ones at the grain boundary, subgrains are also formed in the ferrite crystals, as a result of ATMCRP. Figure 3.15a shows the microstructure of the steel superplastically deformed at 800°C, with bands of ferrite and pearlite, but at a close up (Figure 3.15b and Figure 3.16) each ferritic zone is formed by subgrains; the ferritic boundary can be identified by the presence of cementite precipitates, while the subgrain boundary does not show these structures.

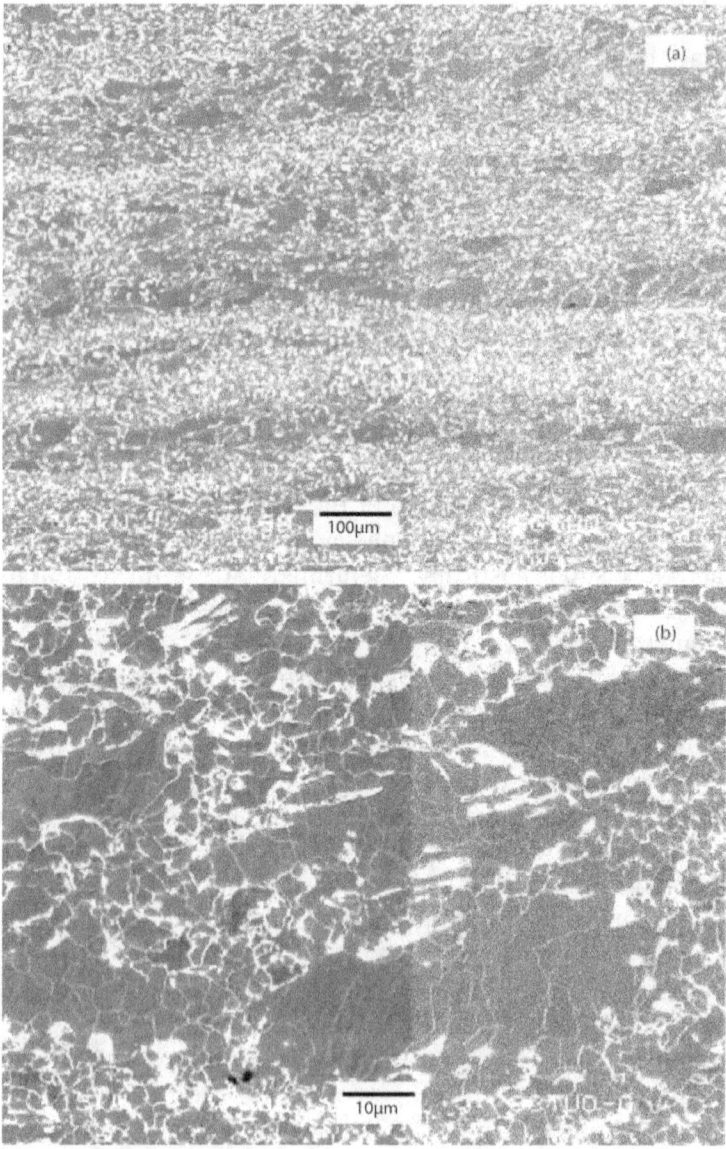

Figure 3.15 SEM micrograph showing bands made of ferrite and pearlite (a) and the subgrains forming the ferrite zones (b)

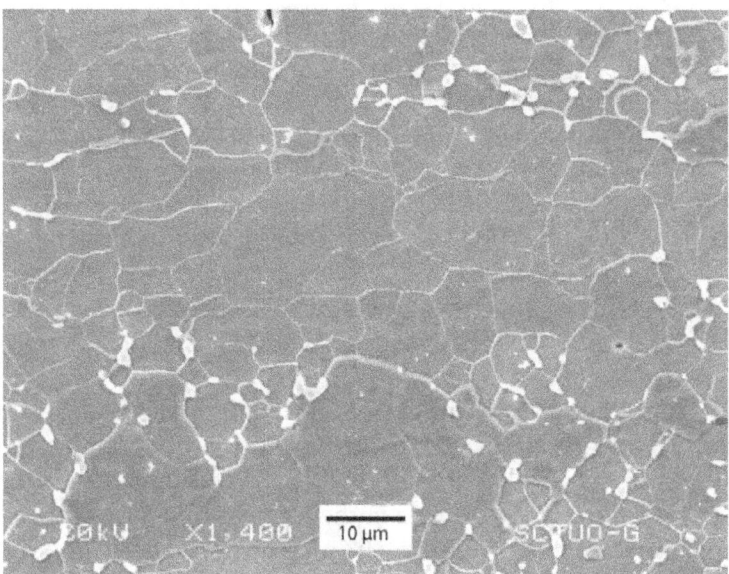

Figure 3.16 Subgrains forming a ferrite zone with cementite precipitates at the boundary

The microstructure of the steel is formed by bands of pearlite and ferrite, which is typical in construction steels that suffered a peritectic reaction and solidification under non-equilibrium conditions. As the proportion coefficient for carbon, alloying elements (Mn, Si, Cr and Mo) and impurities (P and S) in this steel is lower than one, the microstructure cannot be regenerated through soaking treatments before hot rolling processes[34].

The microstructure after high temperature superplastic deformation is a homogeneous one formed by ferrite zones that are slightly elongated in the rolling direction with evidence (subgrains) of having suffered dynamic recovery during deformation along with very fine ferrite grains close to pearlite grains. Furthermore, the following can also be observed:

- The banding has nearly disappeared but there still are zones with large amounts of both ferrite and pearlite.
- Decohesions shaped as w and r, mainly located in the ferrite/pearlite interphase, are unequivocal

proof of intergranular sliding during the deformation process.

- Small cavities in the α-pearlite (previous austenite grains) interphase, which shows different deformation capacity for each of these phases.
- Null evidence of generalized grain growth during deformation, as grain size is very similar to the original one.
- Grain (or grain clusters) sliding and rotating, as a consequence of superplastic deformation.

At 750°C the material does not behave superplastically as the banded microstructure from the raw hot rolled state disappears and only ferrite grains that have suffered dynamic recovery along with carbides in their grain boundaries remain. Furthermore, the theoretical Iron-Cementite Phase Diagram shows that at both 800 and 750°C, the microstructure should be the same. A possible explanation is that these tests imply a negative pressure on the sample (typical phase diagrams consider a pressure of 1 atm) which could induce a change in the eutectoid temperature and its movements upwards in the diagram which would result that at 750°C the material would be on a different zone than at 800°C. In an invariant phase transformation, the temperature variation caused by a pressure change equals enthalpy change divided by the product of temperature and volume change which is known as the Clausius-Clapeyron equation; since austenite has a smaller molar volume than pearlite, then the volume change is less than zero, while the enthalpy change is also less than zero, therefore the temperature variation caused by a pressure change is negative. Furthermore, an increase in pressure lowers the equilibrium transition temperature and viceversa[35].

3.4 Ashby-Verrall model

An HSLA construction steel microalloyed with Ti/Nb (in the form of carbonitrides of Ti, Nb and V) as a sheet with a thickness of 27 mm and manufactured by hot controlled rolling techniques has an ultrafine grain size (12 ASTM G, 5 μm). A few of its mechanical properties are: a yield stress of 447 MPa, high sub-zero tenacity, low cost and easy to weld. This type of materials when produced by ATMCRP and deformed at high temperatures (750~800°C) in the intercritic region $(\alpha + \gamma)$, will present superplastic behavior in tension with elongations higher than 100%.

In the "grain boundary sliding, diffusion-accommodated flow-rate controlling" model proposed by Ashby and Verrall, grain boundary sliding (GBS) with diffusion is considered, subject to maintaining the continuity of a material consisting of a group of four grains each of size $d = 2l$ (where l is the length of a side of the hexagon), and passing through three stages as shown in Figure 3.17. This model is based on three assumptions[21]:

- There is mass transport (by diffusion).
- There is relative movement of the grain boundaries accommodated by diffusion.
- There is rotation of the diffusion-accommodated grains.

The relative movement of the group of four grains shown in Figure 3.17 causes a change in shape of the group as a whole; meanwhile, the grains experience accommodation tension stresses that keep them together. These stresses (which are different from those experienced by the group as a whole) are caused by diffusion[21].

In a polycrystalline material, there are two different modes of diffusion: bulk diffusion (throughout the grains) and grain boundary diffusion. The latter is more important at intermediate and low temperatures (superplasticity is an example of this type of diffusion). For the accommodation of grains after GBS, both types of diffusion are required at low strain rates. With increasing stress, the accommodation stresses due to diffusion are replaced by a uniform deformation mechanism, leading to the disappearance of superplasticity and allowing conventional hot deformation mechanisms to take over[21].

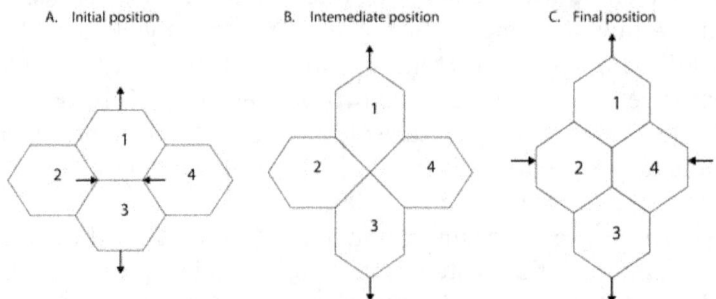

Figure 3.17 Stages of deformation according to the Ashby–Verrall model. Initial (A), Intermediate (B) and Final (C) positions[21]

Considering the initial, intermediate, and final stages of GBS in the group of four grains in Figure 3.17 subjected to tension and constant pressure, it can be seen that the shapes of the individual grains in the initial and final stages are identical, but the geometry of the group has changed (with a real deformation of 0.55). The energy required to plastically deform the material is consumed in the following processes [21]:

- diffusion (mass movement) to facilitate the accommodation (mass continuity) in the grain boundaries;
- GBS, accommodated through the previous diffusion.

Therefore, the Ashby–Verrall equation for superplasticity takes account of two mechanisms, acting in parallel, for the strain rate [21]:

$$\dot{\varepsilon}_{total} = \dot{\varepsilon}_{D\text{-}A\ flow} + \dot{\varepsilon}_{dislocation\ creep}$$

[3.8]

where $\dot{\varepsilon}_{D\text{-}A\ flow}$ (diffusion-accommodated flow) is the strain rate resulting from GBS, accommodated (ensuring continuity of mass) through both bulk and (mainly) intergranular diffusion, which is similar to the Herring–Nabarro mechanism but with two main differences: the first is that the mass volume that needs to be transported by diffusion is lower than in the Herring–Nabarro mechanism and the second is in the threshold stress under which the mechanism would be inoperable. On the other hand, $\dot{\varepsilon}_{dislocation\ creep}$ represents the strain resulting from the plastic flow caused by creep due to the movement and climbing of dislocations, through a mechanism similar to that proposed by Weertman and Dorn[10; 11].

The formulas of the Ashby–Verrall model are:

$$\dot{\varepsilon}_{D\text{-}A\ flow} = 98 \frac{\Omega\mu}{kTd^2} \left(\frac{\sigma}{\mu} - \frac{0.72\,\Gamma}{\mu d}\right) D_v \left(1 + \frac{\pi\delta D_B}{d\ D_v}\right)$$

[3.9]

$$\dot{\varepsilon}_{dislocation\ creep} = A_1 \frac{\mu b}{kT} \left(\frac{\sigma}{\mu}\right)^n \exp\left(-\frac{Q_c}{RT}\right)$$

[3.10]

In order to analyze variation of stress with strain rate, the stress and the shear modulus μ of the material are combined in the parameter σ/μ, allowing determination of the thermal component of the critical shear stress for dislocation sliding. When the strain rate and the parameter σ/μ are plotted, the result is a sigmoidal curve (Figure 3.18), taking the grain size as a parameter, and three stages are found:

- For high and intermediate strain rates, there is sliding and rotation of grain boundaries and the material exhibits typical superplastic behavior.
- For low strain rates, the strain is mainly intragranular, with sliding and climbing of dislocations associated with restoration or

recrystallization processes, depending on the nature of the material.

- The superplastic behavior window can be found in the zone of maximum slope, in other words where m, the coefficient of sensitivity to the strain rate, is higher than 0.3. Grain refinement, as well as increasing the temperature, will improve superplasticity, provided that the microstructure remains stable with increasing strain and the grain size does not grow when the temperature rises, since otherwise there would be a change from dislocation-accommodated creep (necessary for superplasticity to occur) to dislocation creep (conventional deformation behavior).

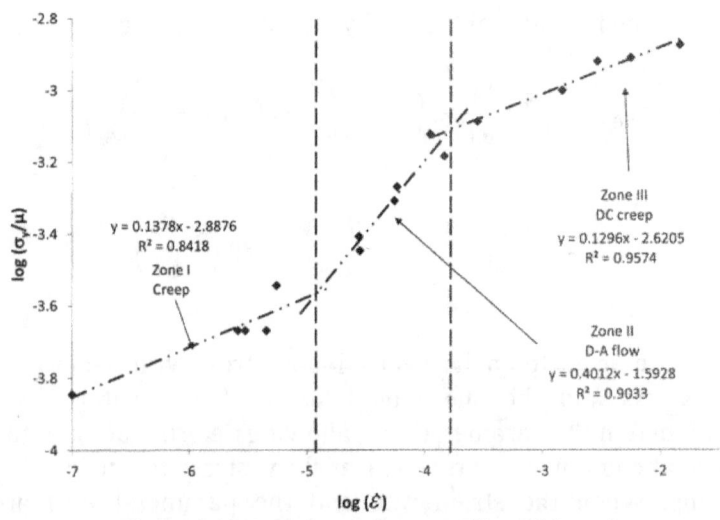

Figure 3.18 Experimental curve of logarithm of yield stress/shear modulus vs. logarithm of strain rate

Superplastic metallic materials are highly sensitive to the strain rate and very insensitive to strain; in other words, the coefficient m has values higher than 0.3, while the coefficient n has values lower than 0.1. Furthermore, the value of m influences the size of the final transverse area of the test specimen: when m is close to 1, the rate of reduction of the transverse area is approximately constant regardless of other parameters. When $m = 1$, the strain is of the Bingham type (Newton viscous flow or $\sigma \approx \dot{\varepsilon}$), which enables any irregularity in the specimen (incipient neck) to be "stable" during the test[18].

The variation of strain rate with strain is positive, in contrast to conventional materials, in which it is negative, producing a "stable" necking condition. Thus, the strain rate varies along the calibrated length of the specimen and is inversely proportional to the transverse section (the Backofen criterion). Moreover, the phenomenon of superplasticity is linked to the nature of the necks, which in this case are spread out along the calibrated length of the specimen in a diffuse way; this is related to the strong influence of the strain rate on the applied tension. When necks are localized, the adjacent areas stop deforming, and stress equilibrium is maintained by the different strain rates belonging to different areas of the specimen[18; 19].

Figure 3.19 shows plots of a fit to the sigmoidal curve describing the yield stress/shear modulus versus strain rate behavior of an HSLA steel according to the following equation:

$$y = -0.0176x^3 - 0.2365x^2 - 0.7605x - 3.6071$$

[3.11]

which has a correlation of 0.9853 (i.e., extremely close agreement between tendency and experimental data). A possible explanation of why the experimental curve crosses the 1, 5, and 10 μm model curves is the grain size distribution present in the steel: there are grains with sizes of 10 μm and grains with sizes of 1 μm. Also, the presence of precipitates of (Ti, Nb)C influences the mechanical behavior of the steel by the anchoring effect: small hard particles pin the grain boundary, preventing grain growth[36].

The modeling of the grain size evolution during the roughing process, waiting time and finishing in construction steels microalloyed with Ti/Nb according to the Sellars – Urcola[37; 38] model, lead to the creation, both in plate (~25mm) as in strip (~5mm), of austenite partially or totally deformed at the end of the ATMCRP. This will result, after the allotropic transformation, in ultrafine ferrite steels (ferrite grain size equal or higher than 12 ASTM G). The banded structure (ferrite-pearlite) phenomenon is directly proportional to the carbon content (0.06~0.17% C).

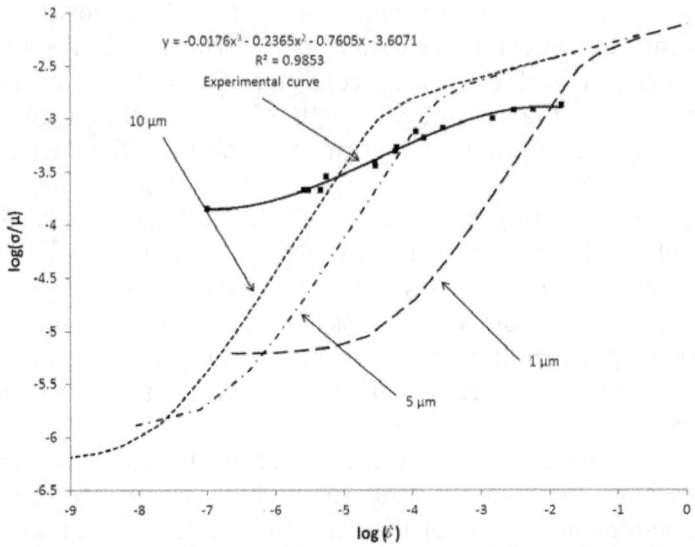

Figure 3.19 Logarithm of yield stress/shear modulus vs. logarithm of strain rate of theoretical curves of different grain sizes (Asbhy-Verral model) and experimental data for an HSLA steel

The mechanism described by Ashby-Verrall[21] seems to be the most appropriate model to describe this superplastic behavior: rotation and sliding of grains activated by diffusion processes at the grain boundaries. At low strain rates ($\sim 10^{-4}$ s^{-1}) the dominating phenomena are rotation and intergranular sliding of the ferrite, diminishing the defects (cavities, pores, decohesions) produced by grain boundary diffusion. At higher strain rates, the mechanism that rules is the dynamic recovery of the ferrite by sliding and climbing of the dislocations (intergranular mechanism) common on conventional hot deformation processes. At the temperature proposed (800°C), inside the intercritic stable ferrite and austenite interval, the construction steels with UFG/UFF microstructure and bands created by hot rolling, with thicknesses lower than 5 mm, would be superplastic at strain rates that would make its conventional manufacturing processes competitive.

When a steel presents superplastic behavior it maintains its fine grain size during the straining; the austenite present (transformed to pearlite during cooling) prevents the growth of the superplastic ferritic grains. The original ferritic-pearlitic banded structure evolves into a more homogeneous structure formed by ferrite and pearlite during straining: the bands blur or disappear because of the non-homogeneous deformation of ferritic grain clusters (sliding and rotating). Though these steels may show superplastic deformation at high temperature, this type of deformation does not imply a generalized grain coarsening of the microstructure even when the strain rate is very low ($10^{-5}s^{-1}$).

Finally, at temperatures lower than 800°C (for example at 750°C) steels with mainly ferritic structure behave differently than at this temperature: the austenite becomes thermodynamically unstable in tension, transforming into ferrite and carbides (mainly cementite); during cooling the ferritic grains grow and the steel would, at least partially, lose its superplastic behavior during slow straining deformation. It is important to point out that after testing at 800°C the microstructure is formed by pearlite, ferrite and precipitates, while after 750°C tests, the room temperature microstructure is only formed by ferrite and precipitates (Figure 3.8c). Therefore the selection of the adequate temperature for straining is critical in UFG construction steels, just as it occurs in conventional superplastic metallic alloys.

3.5 References

1. **Pero-Sanz Elors, J.A.** *Ciencia e ingeniería de materiales: Estructura, transformaciones, propiedades y selección.* [ed.] CIE Dossat. 5th. 2006. (in spanish).

2. *Structural Superplasticity at Higher Strain Rates of Hypereutectoid Fe-5.5Al-1Sn-1Cr-1.3C Steel.* **Frommeyer, G. and Jiménez, J.A.** February 2005, Metallurgical and Materials Transactions A, Vol. 36A, pp. 295-300.

3. *Superplasticity: A Review.* **Davies, G.J., Edington, J.W. and Cutler, C.P.** 1970, Journal of Materials Science, Vol. 5, pp. 1091-1102.

4. *Superplasticity: Prerequisites and Phenomenology.* **Wadsworth, J., Oyama, T. and Sherby, O.** 1980. InterAmerican Conference on Materials Technology.

5. *A composite model for superplasticity.* **Baudelet, B. y Lian, J.** 1995, Journal of Materials Science, Vol. 30, págs. 1977-1981.

6. **Askeland, D.** *The Science and Engineering of Materials.* 3th. s.l. : PWS Publishing Company, 1998.

7. **Reed-Hill, R.E.** *Creep. Physical Metallurgy Principles.* 2nd. Independence, KY : Cengage Learning, 1994.

8. **Ashby, M.F. and Jones, D.R.H.** *Engineering Materials 1.* s.l. : Pergamon Press, 1986. Vol. 34.

9. **Dieter, G.E.** *Mechanical Metallurgy.* s.l. : McGraw-Hill, 1981.

10. *Theory of Steady-State Creep Based on Dislocation Climb.* **Weertman, J.** 1955, Journal of Applied Physics, Vol. 26, p. 1213.

11. *Some Fundamental Experiments on High Temperature Creep.* **Dorn, J.E.** 1954, Journal of the Mechanics and Physics of Solids, Vol. 3, pp. 85-116.

12. *Effect of change of scale on sintering phenomena.* **Herring, C.** 1950, Journal of Applied Physics, Vol. 21, pp. 301-303.

13. *Report on a Conference on the strength of materials.* **Nabarro, F.R.N.** 1948, The Physical Society, p. 75.

14. *Microstructural Aspects of Superplasticity.* **Edington, J.W.** May 1982, Metallurgical and Materials Transactions A, Vol. 13A, pp. 803-715.

15. *An investigation of grain boundary sliding in superplasticity at high elongations.* **Lin, Z.R., Chokshi, A.H. and Langdon, T.G.** 1988, Journal of Materials Science, Vol. 23, pp. 2712-2722.

16. *Principles of superplasticity in ultrafine-grained materials.* **Kawasaki, M. and Langdon, T.G.** 2007, J Mater Sci, Vol. 42, pp. 1782-1796.

17. *Grain boundary engineering for superplasticity in steels.* **Furuhara, T. and Maki, T.** 2005, Journal of Materials Science, Vol. 40, pp. 919-926.
18. *The Elongation of Superplastic Alloys.* **Morrison, W.B.** 1968, Trans. Metall. Soc. AIME, Vol. 239, p. 710.
19. *Superplasticity in an Al-Zn Alloy.* **Backofen, W.A., Turner, R.I. and Avery, D.H.** 1964, Transactions of the ASM, Vol. 57, p. 980.
20. *Superplasticity: Mechanisms and Applications.* **Vetrano, J.S.** March 2001, JOM, p. 22.
21. *Diffusion-accomodated flow and superplasticity.* **Ashby, M.F. and Verrall, R.A.** 1973, Acta Metallurgica, Vol. 21, pp. 149-163.
22. **Alden, T.H.** *Review topics in Superplasticity, Plastic deformation of materials.* [ed.] R.J. Arsenault. New York: Academic Press, 1975. pp. 225-266.
23. *Superplasticity of a Stainless Steel Clad Ultrahigh Carbon Steel.* **Daehn, G.S., Kum, D.W. and Sherby, O.D.** December 1986, Metallurgical and Materials Transactions A, Vol. 17A, pp. 2295-2298.
24. *Superplasticity: A Review.* **Davies, G.J., Edington, J.W. and Cutler, C.P.** 8, 1970, Journal of Materials Science, Vol. 5, pp. 1091-1102.
25. *A microscopic examination of void formation in superplastic materials.* **Langdon, T.G.** May 1979, Journal of Microscopy, Vol. 116, pp. 47-54.
26. *An Analysis of Cavity Growth During Superplasticity.* **Miller, D.A. and Langdon, T.G.** December 1979, Metallurgical Transactions A, Vol. 10A, pp. 1869-1874.
27. *Creep cavitation without a vacancy flux.* **Hancock, J.W.** 1976, Metal Sci, Vol. 10, p. 319.
28. *Vacancy potential and void growth on grain boundaries.* **Speight, M.V. and Beeré, W.** 1975, Metal Sci., Vol. 9, p. 190.
29. *Diffusional growth of creep voids.* **Harris, J.E.** 1978, Metal Sci., Vol. 12, p. 321.
30. *Diffusion along Grain Boundaries with Non-Equilibrium Structure.* **Valiev, R.Z., Razumovskii, I.M. and Sergeev, V.I.** 1993, phys. stat. sol., Vol. 139, pp. 321-335.

31. *Grain Boundaries during Superplastic Deformation.* **Valiev, R.Z., Kaibyshev, O.A. and Khannanov, S.K.** 1979, phys. stat. sol., Vol. 52, pp. 447-453.

32. *Influence of Thermo-Mechanical Processing Parameters and Chemical Composition on Bake Hardening Ability of Hot Rolled Martensitic Steels.* **Asadi, M. and Palkowski, H.** 7, 2009, Steel Research International, Vol. 80, pp. 499-506.

33. *Effect of Composition and Process Variables on Nb (C,N) Precipitation in Niobium Microalloyed Austenite.* **Dutta, B. and Sellars, C.M.** 3, 1987, Materials Science and Technology, Vol. 3, p. 197.

34. **Pero-Sanz, J.A.** *Ciencia e Ingeniería de Materiales.* Madrid : CIE-Dossat 2000, 2006.

35. **Porter, D.A. and Easterling, K.E.** *Phase Transformations in Metals and Alloys.* s.l. : Van Nostrand Reinhold, 1988.

36. **Porter, D.A., Easterling, K.E. and Sherif, M.Y.** Phase Transformations in Metals and Alloys. 2009, 2, pp. 63-110.

37. *Recrystallization and grain growth in hot rolling.* **Sellars, C.M. and Whiteman, J.A.** 1979, Metal Science, pp. 187-194.

38. **Urcola, J.J. and Fuentes, M.** 6, 1980, Revista de Metalurgia del CENIM, Vol. 16, pp. 337-342.

About the Authors

María José Quintana, PhD (mquintana@up.edu.mx) has an European Doctorate in Science and Technology of Materials from the University of Oviedo (Spain), and is a professor and researcher at the Faculty of Engineering at Panamerican University (Mexico). Works on subjects of molding, thermomechanical treatments of steels, superplasticity of metals, microscopy and mechanical properties of materials, among others.

Roberto González (robglez@up.edu.mx) has a PhD in Materials Technology from the University of Navarra (Spain) and twenty years' experience in research related to the automotive industry and design methodology practices. He is currently researcher and professor at the Faculty of Engineering at Panamerican University, Mexico City and his interests include the design and characterization of metallic materials, finite element simulation of mechatronic systems and tribology (wear phenomena) analysis of industrial facilities.